OPTICAL ANECDOTES

Ibn al-Haitham's experiment with the camera obscura

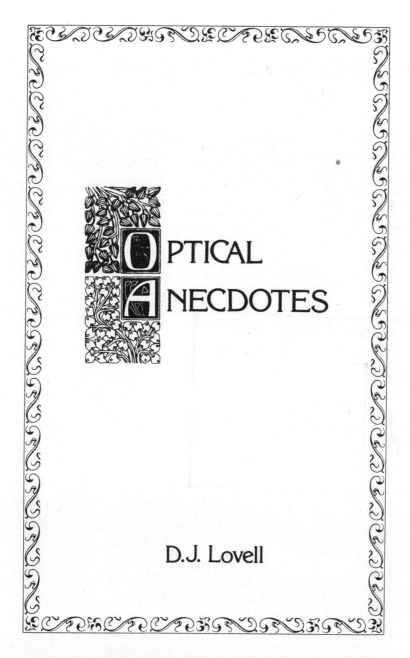

OPTICAL ANECDOTES

D.J. Lovell

SPIE – THE INTERNATIONAL SOCIETY FOR OPTICAL ENGINEERING
POST OFFICE BOX 10   •   BELLINGHAM, WASHINGTON 98227 USA

Library of Congress Catalog Card No. 81:52767

ISBN 0-89252-353-0

Printed in the United States of America. Fourth printing July 1984.

To the honor of those investigators who have provided us with the enjoyment attained by our understanding of the nature of light.

Contents

Contents

Illustrations

Illustrations

Preface

NDOUBTEDLY, NEARLY ALL WHO READ THIS BOOK HAVE, at one time or another, pondered the question, "What is light?" To answer, "Light is that which permits vision," begs the question, for such an answer provides us with no understanding of the nature of light. It says no more than "light is light."

Describing the nature of light has been bothersome for countless years. This is attributable to a variety of facts. Most important, light is a subtle matter that is perceived by imperfectly understood sensors. (Does the light reflected from this page induce a perception that is markedly different from the sensation achieved in a dream?) Light is also manifest in an exceedingly broad range of circumstances. Light is emitted from hot bodies, from lightning discharges, from some insects, and from some not well-defined processes. Light is all about us, stimulating our most perceptive sense—vision.

Light is also of great aesthetic value. The poet, William Wordsworth, wrote, *My heart leaps up when I behold/ A rainbow in the sky*. This thought is reflected in the emotion occasioned in the minds of countless men from the dawn of antiquity. A study of the rainbow evokes added joy to its perception. Such a study entails gaining some knowledge of the nature of light.

From this deeper appreciation of the rainbow we are led to study such fascinating phenomena as halos, mirages, and phantasmagoria. Simpler things, too, command our attention: the blue of the sky, the red of the setting sun, the whiteness of clouds, blue shadows on snow-covered fields, and the colors of flowers and birds. Light is certainly fundamental in much of our aesthetic pleasure. Apart from these pleasures, we study the nature of light because of its role in technology. The telescope and microscope, photography, lasers, optical fibers, and many other devices are based upon a knowledge of the nature of light.

No wonder, then, that man asks, "What is light?" We frequently resort to operational definitions in an attempt to answer that question. That is, when we confront such phenomena as the photoelectric effect we conceive of light as a bombardment of miniature bullets, or corpuscles. On the other hand, light is evident by such phenomena as diffraction, polarization, and interference. This leads us to conceive of light as causing a wave disturbance. This dichotomy was resolved by quantum mechanics from which these macro-analogies were replaced with a conception of light existing in atomic dimensions with sufficient momentum to interact with matter and produce the photoelectric effect, but guided through space by an electromagnetic wave.

The history of man's investigations of the nature of light is fascinating. It encompasses a large cast of characters who interact with all facets of the history of mankind. The anecdotes presented here describe a few of the events which took place in man's search for the nature of light—a search which included drama, adventure, humor, political and religious entanglements, pestilence, and the many foibles of men. Some of the men involved displayed enormous conceit; others were humble. Some were born to wealth; others had to overcome poverty.

I will consider my task accomplished if some who read this are motivated to a greater joy in observing nature's splendors and to extend themselves to further investigations of the nature of light. You must not, however, consider these anecdotes to be more than stories informing you of some of the incidents in the lives of the men who made significant contributions to optics. This book is neither a history nor an optics text.

The material used in preparing these anecdotes was derived from various sources. The books which dominated the search are listed in the bibliography. In addition, journal articles have proved to be an invaluable source of information. Most joyful of all the sources, undoubtedly, were visits to museums. Those in Europe are of particular value for historical research. Not the least of my sources has been numerous personal interviews. Because my sources are so diverse, I have refrained from citing references. To do so would obscure the text and, in any event, it would be of marginal benefit to the reader if I were to cite discussions or visits. Naturally, I shall be pleased to assist anyone desiring to locate specific source material.

A few of the anecdotes have been published in approximately their present form in *Optical Engineering* and *ISIS*. The anecdotes constitute distinct stories that need not be read in order. Some attempt has been made to arrange them chronologically but, as this is not intended to be an historical account, the arrangement is not considered of importance.

To cite my appreciation to all who have contributed to this effort would require several pages. Suffice it to say that I have received considerable help for which I am very grateful.

I've enjoyed preparing these anecdotes; I hope that you enjoy reading them.

D. J. LOVELL

40 Barton Road
Stow, Massachusetts 01775
July 1981

Introduction

HE SECOND WORLD WAR interrupted D. J. Lovell's education at the University of Wisconsin, but contributed to his experience in optics because his hitch in the Navy included two years at M.I.T. working in Arthur Hardy's research group. War over and Masters Degree in hand, D.J. started his professional career teaching at Norwich University in Vermont. The good pay, $2700/year, did not prevent him from seeking greater use of his skills; he accepted a Civil Service job at the Naval Research Laboratory in Washington. That was in 1949. D.J. joined the Radiometry Branch of the Optics Division of NRL.

At that time, I was forming a group within the Branch to do several optical experiments for the Los Alamos Scientific Laboratory at the upcoming tests of nuclear devices to be conducted at Eniwetok Atoll. The Branch Head, John Sanderson, suggested that D.J. might serve in the group.

There must have been some sort of interview or initial encounter in which the details of education and experience were discussed. I have no memory of this. I do remember knowing from the outset that the group had gained a man who, after high school graduation in 1939, had paddled the length of the Mississippi River from Lake Itasca to New Orleans and, in the process, had lost much of the photographic record when his canoe overturned near Little Rock.

During the more than thirty years since then, our relationship hasn't improved particularly. The fault is mine, I know. I continue to delight in D.J.'s adventures in optics and sometimes overlook his more serious contributions to the discipline. I am comfortable with this flaw in my nature, now that some of D.J.'s essays are available in one volume. The emphasis of the book somehow reminds me of the emphasis of his career. In the essays he treats the great moments and the great men of optics with careful attention to the unusual and often little-known dramas of personal life and history which surround the celebrated achievements of greatness. He believes that the achievements themselves are adequately treated by the more sedate textbooks of our trade.

It seems to me that his collection of anecdotes can be pleasant reading for each of us, can be a useful source book for teachers, and can be a delightful reference for students. Where else can one find in one volume the story of Arago's encounter with an engaged French woman, the reason why Tycho had a nose of silver and gold, how Brewster invented the kaleidoscope, and many other entertaining sidelights to optical history?

There are sidelights to D.J.'s career which in themselves could be the subjects of a collection of anecdotes. Fortunately, D.J. loves to tell a good yarn. One day he may put such stories on paper. Until then, should you meet him, ask about his color photographs of the Green Flash made over thirty years ago at Dry Tortugas, about being low man on the totem pole during the eclipse expedition to Khartoum, about an emergency helicopter landing on the lip of a still-smoldering crater formed by a nuclear explosion, about—well—about almost anything, or just ask him to tell you about himself.

HAROLD S. STEWART

North Andover, Massachusetts

Optics in Antiquity

The Pyramids, Stonehenge, Aristophanes, The Egyptians, The Mesopotamians, The Greeks,
The Hellenists, Plato, Aristotle, Alexander the Great, Euclid, Hero, Seneca, Ptolemy

 S MAN EMERGED FROM HIS PRIMEVAL STATE, he became conscious of the formation of shadows, the color of the daytime sky, the
spectacle of the rainbow, and the motions of the stars at night and the
sun in the day. Thus, he began his awareness of light. Initially, he could
seek to discern the nature of light only by noting its rectilinear motion. But this soon
led to observations which tended to clarify the understanding of mechanisms responsible for these phenomena.

¶ Early men, equipped with little more than a stick mounted vertically in the ground,
were able to discern the rotation of the earth and its precession by observing the shadow cast. Many structures have been found which attest to this. The pyramids of Egypt
and the massive stone circle at Stonehenge are examples. Once thought to be no more
than religious edifices or tombs of vainglorious pharohs, they are now seen to reflect
early man's awareness of terrestrial motion in the solar system.

¶ The great pyramid at Cheops is so finely oriented that compasses can be adjusted
by reference to it. Some believe the pyramids may be a scale model of the earth's
hemisphere which served as an almanac by means of which the year could be measured
to within about a minute of time. And this from a structure that rises to a height of a
modern forty-story building with a base that covers some forty acres! This pyramid
contains more stone than all the cathedrals, churches, and chapels built in England
since the time of Christ.

¶ The circle of sarsen stones at Stonehenge in Wiltshire, England, also suggests that
the ancients knew considerable about the motions of celestial bodies. The alignment
of these stones with significant astronomical events, such as the rising of the sun at the
summer solstice, cannot be interpreted as mere happenstance. It has been suggested
that these stones were arranged to constitute a neolithic computer to predict seasons,
eclipses, and similar phenomena.

¶ The engineering feats associated with both the pyramids and Stonehenge are im-
pressive. The massive stones at Stonehenge were transported over hundreds of kilometers and were positioned with extreme accuracy. In a like manner, the stones from
which the pyramids were built were carried over an extreme distance and positioned
with precision. I, who have difficulty in positioning a sundial to give approximately correct time, marvel at the abilities displayed in Egypt and England over 4,000 years ago.

¶ The first recorded account of the use of lenses was not written until 425 B.C. by
Aristophanes in his *The Clouds*. It is quite conceivable that earlier accounts were written and subsequently lost, since archeological findings of lenses suggest they were
known 3,000 to 3,500 years ago. In any event, Aristophanes, possibly with tongue in
cheek, suggested that a burning glass could be used to destroy *at a distance* the writing

formed on a wax tablet, thereby giving an advantage to anyone confronted by a bailiff serving a summons.

¶ All branches of science underwent interrupted growth in early times. Achievements attained by one element of society may not have been continued by another. Advances were accomplished, distinction was achieved, and then that society faded. The transfer

These sarsen stones at Stonehenge, erected some 4,000 years ago, are aligned with significant astronomical events, suggesting that they were arranged to constitute a neolithic computer to predict seasons and eclipses.

of knowledge between contemporary societies or to subsequent generations was certainly not done astutely 3,000 years ago. The Egyptians and the Mesopotamians from about 3500 to 300 B.C. achieved sporadic gains. Some of this was learned by the Greeks beginning about 600 B.C., but within a millennium their remarkable advances were silenced. However, the impact they made influences much of science today.

¶ Several characteristics of light were recognized by the Hellenists and the nature of light began to be 'understood,' although the Greeks seem to have been more concerned with vision than with the nature of light.

¶ Plato (427 to 347 B.C.), who had been a student of Socrates, taught that vision is achieved by the expulsion from the eye of *ocular beams* that are propagated with great speed in straight lines. These rays cause sight when they meet with something that emanates from the object. One may question today how anything so seemingly inconceivable could have been advocated by this great philosopher. I suspect that the basis for the adoption of the ocular beam lies in the regard the Greeks had for the soul, from which they believed the senses emanated. Thus, sensation would be derived from within the body although the cause would be external. Even today, our speech includes such phrases as "cast your eyes on this" and "such-and-such is at the limit of sight."

¶ Aristotle (384 to 322 B.C.) studied under Plato but rejected the concept of ocular beams, arguing instead that vision arises when particles emitted from an object enter the pupil of the eye. Aristotle is best remembered for his encyclopedic classification of existing knowledge. For the following eighteen centuries, the philosophic basis of his science went practically unchallenged. In optics, he ascribed twilight to the reflection of sunlight by the air surrounding a spherical earth. He attributed the colors of the rainbow to reflection of sunlight from drops of rain.

¶ We thus see men beginning to investigate the nature of light itself. Aristotle's theory of vision was adopted by the Pythagoreans, Democritus, and others. But the concept of ocular beams was, by no means, put to rest.

¶ Hellenistic science reached maturity about the time of Alexander the Great (356 to 323 B.C.). Although he had been a student of Aristotle, he did not take up philosophic studies as a profession. Rather, through military conquest he carried Greek culture to the civilized world and initiated a new period in history which lasted for about six centuries. During this period, several philosophers were enabled to make contributions to our understanding of the nature of light.

¶ Euclid (c. 300 B.C.) lived in Alexandria shortly after the great conqueror died. Euclid's greatest fame, of course, is derived from his *Elements of Geometry* which, in substance, is still studied in our grammar schools. Little is recorded of his personal life, which is consistent with our present-day inclination to favor the commemoration of the lives of politicians, military leaders, athletes, movie stars, and wealthy men. However, Euclid made important contributions to our interest in light with his *Optics* and *Catoptrics*, although there is some question whether he was the author of the latter.

¶ Euclid advocated Plato's concept of ocular beams and, upon this basis, he constructed a theory of perspective relating qualitatively the relationship of objects seen in space. He also described the formation of images by spherical and parabolic mirrors. From this, he stated that a concave mirror turned toward the sun will cause ignition in flammable materials placed in the focal plane. These observations were similar to those of the effects of spherical pieces of glass made in antiquity.

¶ By the beginning of the Christian era, several significant contributions to optics were being made by Greek savants. Hero of Alexandria (c. A.D. 60), for instance, made several observations leading to an improved understanding of several facets of science. We are particularly concerned with his *Catoptrics* in which he describes the reflection of light from mirrors, demonstrating that the angles of incidence and reflection are equal. Of greater significance, he showed that this results from the principle that light travels by the shortest distance from its source to the observer's eye. This is particularly relevant as it results from a deduction made concerning observational data.

¶ Seneca (4 B.C. to A.D. 65) observed white light dispersed by a prism. He deduced that the colors observed were fictitious, arriving at this conclusion by comparing the colors with the iridescent appearance of certain bird feathers which cannot be attributed to pigments in the plumage. Seneca's reasoning was thus not entirely correct and did little to explain the nature of light. However, in all phases of science, suggestions are made which are faulty but which stimulate further enquiry.

¶ In my (biased) opinion, the greatest of the Hellenistic scientists was Ptolemy (c. A.D. 150) of Alexandria. His *Almagest* and *Geography* dominated astronomy and geography for fourteen centuries. He listed basic observations of the earth and solar system. He established terms equivalent to latitude and longitude and gave this information for the important places of that time, thus forming an atlas of the then-known world. He understood both orthogonal and stereographic projections of the sphere. Ptolemy prepared the earliest star catalog that has survived. He determined the distances of the sun and the moon from the earth, and he predicted eclipses.

¶ In his book *Optics*, Ptolemy suggested that the eye emits rays similar to those described by Plato. However, he argued that the rays are continuous rather than consisting of minute particles. Ptolemy studied reflection and refraction and showed that the incident ray, the reflected ray, and the normal to the reflecting surface all lie in the same plane. He obtained similar results from studies of refraction. He knew that the angle of refraction does not equal the angle of incidence, and he published three refraction tables—for light passing from air to water, from air to glass, and from water to glass.

¶ Ptolemy is sometimes known as the last astronomer of antiquity. By this time, the size of the earth had been measured. The sizes of the moon and the sun were reasonably known, as were their distances from the earth. The precession of the equinoxes and the length of the year had been established.

¶ Studies of the geometrical characteristics of the passage of light through various media had had a good beginning in ancient Greece. However, the physical nature of light had not been described, and the physiological aspects of vision then proposed seem somewhat strange today.

¶ Following this period of flourishing culture, there was a sad decline. Attila, King of the Huns, took power in A.D. 433 and was responsible for the destruction of over seventy prospering cities. Rome, which had inherited the Greek culture, suffered from an internal cancer that rotted its economic, social, and moral structure. European scientific pursuits stagnated into a Dark Age. Fortunately, the knowledge amassed was preserved for the Renaissance.

The Greatest Authority on Optics in the Middle Ages—Ibn al-Haitham

OUR COMPREHENSION OF SCIENTIFIC ACHIEVEMENTS is often limited to those developments recorded in one's own country or in those allied to it. That this is so is probably attributable to patriotic chauvinism and religious prejudice. For instance, we are likely to consider that the optics we practice (as well as many other cultural activities) began in Greece and came to us directly by way of Europe. Let us consider the Arabian influence on optics and dispel some of this myth.

¶ About A.D. 575 in the Arabic city of Mecca a humble boy was born, who came to be known as Mohammed (which translates roughly as *highly praised*). He founded the Muslim faith and revitalized the pursuit of culture in the Arabic countries. In this way, the Arabs became heir to Hellenistic science, since they had been part of the empire founded by Alexander the Great. Fortunately, they not only inherited this knowledge but they nurtured it, while European culture stagnated through the Dark Ages.

¶ The difficulty of depicting the advances experienced by the Arabs is attributable not only to the prejudices cited above, but also to the fact that most of the original material has been lost. Nevertheless, sufficient evidence is extant to ascertain considerable about the developments which took place. Following Mohammed's ascendancy, the initial scientific effort consisted primarily of translating the Greek and other sources into Arabic. However, numerous advances in science were being recorded. Probably the first of those who made substantial contributions was the philosopher Abu Ysuf Yaqub Ibn Is-haq (813 to 873), better known in the Western world by the Latinized name Al-Kindi. Born in Kufa, he later lived in Basra and Baghdad where his efforts ranged over broad aspects of physics. Al-Kindi wrote *De Aspectibus* in which he dealt explicitly with optical problems. He asserted that vision had to take place by means of rays which promote a physical reaction upon the eye. This was in contradiction to the hypothesis advocated by Plato, which considered that vision is accomplished by the expulsion of *ocular beams* from the eye. Al-Kindi attacked this concept, arguing that ocular beams are but mathematical abstractions which are incapable of acting physically or physiologically. Al-Kindi extended the concept of a visual ray, noting that the formation of shadows suggests that light travels in straight lines. However, he was unable to explain how the rays could react within the eye to send necessary information to the soul permitting a visual reconstruction of the physical world. The concept of ocular beams thus remained widely accepted.

¶ By the time of Al-Kindi's death, Arabian science was flourishing. Into this activity in A.H. 354 (A.D. 965) in Basra was born Abu Ali Mohammed Ibn al-Hasan Ibn al-Haitham, generally known in Arabian countries as Ibn al-Haitham and in Western countries by the Latinized Alhazen. I shall refer to him here as Ibn al-Haitham. Although not much is known of his parents or of his early life, it can be inferred that he belonged to a middle-class family which was sufficiently well-to-do to provide him with an education, but not rich enough to provide him with the leisure to seek higher learning. He secured a position in a government office, suggesting that he must have done well in his studies and that his family exerted some local influence.

¶ The government position provided al-Haitham with means for subsistence, but did not provide any intellectual stimulation. That was acquired during his spare hours by studying astronomy, mathematics, physics, and medicine. As his knowledge and self-confidence grew, he sought new horizons. He was naturally attracted to Egypt where the ruler, Fatimid Caliph Al-Hakim, was a patron of learning and had drawn several scholars to his court. Ibn al-Haitham reasoned that Al-Hakim would require some reason to invite him to Cairo. Thus, he made a thorough study of the Nile, which was and is the life blood of Egypt. The people of Egypt depend upon the water for irrigation, but dread the annual flood which causes considerable damage. Ibn al-Haitham concluded that a dam would entrap the water for the dry season and prevent the annual floods. He then prepared the outline of a plan to build such a dam at a site near Aswan, and sent it to the Fatimid Caliph Al-Hakim. Naturally, Al-Hakim was impressed and invited Ibn al-Haitham to his court. The date at which this occurred is uncertain, even though, as we shall see, it is somewhat relevant. Al-Hakim was only eleven years old when he ascended the throne in 996. It seems unlikely that he would have achieved the maturity to invite al-Haitham to his court until perhaps a decade later. When he arrived in Cairo, then, al-Haitham must have been at least in his early forties.

¶ One of the first things that al-Haitham did upon his arrival was to visit Aswan. Prior to this, he had not even visited Egypt, having made all of his plans for the dam from maps and studies available of the geography and geology of Egypt. With adequate funding, al-Haitham traveled to Aswan, some 400 miles from Cairo. Here he made detailed surveys of the topography of the area, sampled the soil, noted the formation of rocks, and measured the width and discharge of the river. The Nile, incidentally, discharges about 17,000 cubic feet per second in the dry season and over 300,000 cubic feet per second after the rains. Ibn al-Haitham soon realized that the task of constructing a dam was beyond the engineering capabilities of the time. This left two courses available to him. Either he could dilly-dally with the project, thus delaying the day of reckoning, or he could make a clean breast of the situation and hope for the best, knowing that the Caliph was short-tempered and inclined to be cruel. Being a man of high integrity, al-Haitham chose the latter route. Fortunately, Al-Hakim suppressed his anger and passed no harsh order. But his subsequent behavior made it abundantly clear that Ibn al-Haitham had lost his favor and that the Caliph held a strong grudge against him. To avoid a harsher treatment, al-Haitham feigned madness. This was done so well that the Caliph was persuaded only to imprison al-Haitham and to confiscate all his books and scientific instruments. Again, good fortune smiled upon optics, for Al-Hakim died in 1021, and al-Haitham was realeased. Since it is uncertain when he was imprisoned, we do not know how long he was deprived of his books and instruments, but certainly for an extended period al-Haitham had little to do but contemplate. At any rate, upon release he soon settled down to leading a scholar's life. Living near the University of Azhar, he earned sufficient money to meet his needs by copying such books as Euclid's *Elements of Geometry* and Ptolemy's *Almagest*.

¶ Ibn al-Haitham thus plunged into the passion of his life—scientific pursuit. At an age in his mid-fifties, he began valuable contributions to geometrical and physiological optics, which continued for over two decades. The results of his optical researches were recorded in his *Kitab-ul-Manazir* (Treatise on Optics).

¶ This was done during a period when there was still a strong submission to the writings of the ancient scholars. Aristotle was referred to, even by al-Haitham, as "the

This painting depicts the remarkable achievement of Ibn al-Haitham's employment of the camera obscura, experimentally demonstrating that, if an object is placed in a dark room and is irradiated by light passing through a tiny orifice, the image of the object formed on a white screen would be inverted. In June 1981 the Hamdard Foundation, Pakistan, deputized the artist, Zia, to create this picture of al-Haitham's experiment for this book. *(Painting courtesy Hamdard Foundation, Pakistan)*

master" and his work was considered inviolable. However, al-Haitham considered that experimentation was necessary to the understanding of a phenomenon.

¶ One of al-Haitham's first contributions consisted of rejecting Plato's ocular beams. He argued that the apparent sizes of objects viewed at a distance are easier explained by rays of light emitted from the object than by ocular beams.

¶ Through experimentation, al-Haitham divided transparent bodies into two classes: "celestial" and "sub-celestial." The former is absolutely transparent and became the forerunner of the concept of the ether. The sub-celestial bodies were divided into three sub-categories, consisting of gases, fluids, and solids. He maintained that the propagation of light through a transparent body is a physical characteristic of all kinds of light rather than a characteristic of the body.

¶ Ibn al-Haitham studied the phenomena of reflection and refraction, showing experimentally that the incident ray, the reflected (or refracted) ray, and the surface normal all lie in a plane. His method is often used today to illustrate this. He attempted to quantify the law of refraction, but could show only that the ratio of the angle the incident ray makes with the surface normal to that made by the refracted ray is about 1.3 for angles less than twenty degrees. Although Muslim mathematicians had worked out the concept of the sines of angles, it would be about 600 years before the correct law of refraction was stated.

¶ Ibn al-Haitham's studies included the formation of images by spherical and parabolic mirrors. His pioneering efforts in this field were accepted as the standard work for several centuries. He also made a pinhole camera, and described how an inverted image of a candle is formed thereby.

¶ Perhaps his greatest contributions to optics lie in his efforts in physiological optics. He described the various parts of the eye and their function. The opaque coating of the eye outside the iris which forms the white of the eye and is known as the sclerotic was discussed. The function of the horny transparent structure in front of the eyeball, known as the cornea, was studied and reported. He also investigated the characteristics of the membrane behind the cornea (the choroid), the iris, the aqueous humor, and the retina. His description of the eye has been termed masterly and remains as the basis for description today. Several of the terms he used have been translated literally into Latin and are used today. For instance, the lens of the eye suggested to al-Haitham a certain grain, so he used the word *adasa* which is the Arabic word for this grain. This particular grain, lentil, is known as *lens* in Latin, and thus *adasa* was translated to *lens*. It is interesting to note in passing that literal translations sometimes have amusing results. For instance, the English word *airline* translates into an Arabic word suggesting geometrical lines in the sky!

¶ Ibn al-Haitham also described the formation of a halo, the scattering of light by dust particles in a darkened room, the duration of twilight, and similar phenomena.

¶ At the age of seventy-four, al-Haitham died. Although Westerners may be unaware of his life, he has been called *the greatest authority on optics in the Middle Ages*. His work is said to have had a profound influence on Roger Bacon, Wittelo, Leonardo da Vinci, Johann Kepler, and Sir Isaac Newton. And this was accomplished by an Arabian scholar after he was released from prison at an age greater than fifty.

The Evolution of Optical Technology

The Gnomon, The Merkhet, the Meridional Armillary, The Parallactic Instrument, The Clepsydra, The Astrolabe, The Sun Dial, The Necessaire, The Ring Dial, and Spectacles

ECHNOLOGY IS SOMETIMES CONSIDERED to be the handmaiden of science. However one considers their association, it is a close, important, and necessary relationship. Early optical instruments were devised for utilitarian purposes, primarily to aid astronomical observations. These observations, in turn, permitted more accurate measurements of stellar positions to be made, which led to an improved understanding of the passage of light through the atmosphere. From this, it was possible to unlock the mysteries of the rainbow, the colors of the clouds and sky, and other properties of the nature of light. All this took time, of course, and required a continual improvement in optical technology.

¶ Undoubtedly the oldest optical instrument was the simple *gnomon*, which consisted of a post erected upright in the ground. The direction and length of the shadow cast by this device varied according to the time of day and to the season of year. Ptolemy used a gnomon to determine the times of equinoxes and solstices. Probably, the pyramids of Egypt served as a gigantic gnomon. A portable gnomon was made using a hollow hemispheric bowl with a small vertical pointer at the center. Lines were engraved on the inner surface of the bowl to mark the position of the shadow at different hours of the day.

¶ Another simple instrument was the *merkhet*, which consisted of a strip of wood into which a narrow, deep, V-shaped notch was cut at one end. A plumb line was used in conjunction with this to determine the culmination of a particular star. This was accomplished by placing the plumb line at some distance from the V-notched stick and noting the time when the star being observed was viewed exactly in line with the notch and plumb line. Such an instrument was used by the Egyptians some 5,000 years ago.

¶ A more sophisticated instrument serving approximately the same purpose as the merkhet was the *meridional armillary*. This device consisted of a bronze ring, accurately made with a uniform cross section and having a diameter of twenty-five centimeters or more. The ring was clamped onto a fixed pillar. The lateral face of the ring was graduated into 360 degrees, and each degree was subdivided into five minutes. Note that if the ring were fifty centimeters in diameter, a five-minute division would be spaced by about 0.36 millimeters. Inside the fixed ring, a smaller ring was fitted so that it could be rotated in contact with the fixed outer ring. Small plates were fixed at opposite ends of a diameter of this inner ring to serve as sights, so that both the time of conjunction and the altitude could be measured.

¶ Both the meridional armillary and the merkhet were necessarily aligned accurately to the plane of the meridian. This could be determined by measuring the shortest shadow of a gnomon. Although the merkhet could, in principle, be used to determine altitudes, it would be far easier to do so with the armillary. This was used to find the height of the winter and summer solstices.

¶ Another device of interest was the *parallactic instrument* which was used to measure the zenithal distances of stars and the moon at the time of culmination. This, in effect, is the ancestor of the modern transit. It consisted of a vertical bar, over two meters in height. At the top of this was hinged a wooden spar sufficiently thick to be rigid and of about the same length as the bar. At the bottom of the bar was hinged, by a sliding couple, a boom which was also engaged to the free end of the spar. A star was observed by aligning the spar to it. The triangle formed by these three members could then be solved to give the zenithal distance. Oddly enough, Ptolemy was not aware that this determination could be simplified by placing graduated marks on the boom. That was accomplished by an Arabian astronomer in the ninth century.

¶ Optical instruments used for astronomical observations generally do not require that they be portable. Hence, a number of observatories existed in medieval times. These were hardly more than ordinary houses situated in an open position so as to command a good view of the horizon. They were, therefore, located away from metropolitan areas. Note that we still locate observatories in remote areas, but for a different reason.

¶ Aside from its interest to astrologists, astronomy was studied primarily for the purpose of constructing calendars and for determining terrestrial positions (latitude and longitude). Of course, it was necessary to utilize clocks to aid these measurements. The *clepsydra*, or water clock, was probably the basic instrument for the determination of time at night during ancient times. It consisted of a container of water fed by a constant flow. A float rising with the water marked the passage of time. These instruments were probably accurate to within ten minutes. The calendar of Omar Khayyam of 1079 was as accurate as any until the Gregorian reform in 1852.

¶ The determination of latitude was generally accomplished by the use of an *astrolabe*. This, of course, would be of great consequence for a traveler. Determining longitude, on the other hand, also requires a knowledge of time. This can be determined readily by a stationary observer, but presents some difficulties to a traveler.

¶ Although the astrolabe was known to Ptolemy, its widespread use dates to about the tenth century. One of the early descriptions of the astrolabe was written by Geoffrey Chaucer in the fourteenth century. His *Tretis of the Astrolabie* was addressed to *Ltel Lowys my sone* and states, "I nam but a lewd compilatour of the labour of olde Astrologiens."

¶ The astrolabe is a device for measuring the angular distance between two objects. It consists of a graduated circular plate with a datum line and a rotatable pointer called the *alidade* on which are located two sights. The astrolabe is hung from a ring at the top of the diameter perpendicular to the datum line. The alidade is rotated until it points at a particular star, whose altitude can now be read.

¶ However, the instrument does far more than merely determine altitudes. The celestial sphere is inscribed on the side of the astrolabe opposite to that supporting the alidade. This sphere is formed by projecting it from its south pole onto the plane of the equator. Thus, the solstitial circles and the equator are mapped concentrically. The rim of the astrolabe is made to coincide with the winter solstitial circle and is divided into 360 degrees to permit measuring right ascension, i.e., equinoctial time.

¶ Knowing the longitude of the sun (e.g., from a table) permits one to ascertain his

An astrolabe such as this had its origin in Hellenic times, but was highly developed by the Persians. It is the precursor of the modern sextant. *(Photographed by the author at the University of Aberdeen. The instrument was held by the Chief Librarian there.)*

position on earth. This feature of the astrolabe is useful only near the latitude for which it was prepared, but different plates could be substituted for various latitudes, enabling a traveler to use the instrument over prolonged journeys.

¶ It is interesting to note that in addition to its use as a navigational tool and astronomical instrument, the astrolabe could be used to cast horoscopes, to survey, to perform some calculations, and to undertake quite a variety of other tasks. Nevertheless, the determination of time was still an obstacle to its use over long journeys.

¶ Sun dials, of course, can be used to determine time, but these are not portable unless they can be combined with a means of locating north. The compass was not readily available until the end of the fifteenth century. By that time, the *necessaire* came into use. This was a compact device that could be used as an astrolabe, compass, sun dial, quadrant, calendar, calculator, and a moon-phase reckoner. One can imagine the convenience this provided a traveler!

¶ The ring dial is another example of optical instruments in use prior to the Renaissance. It was simply a small ring to which were attached sufficient surfaces to serve as a sun dial at any latitude, providing, of course, that one knew how to locate north. This could be ascertained by locating the North Star, establishing a meridian, and waiting for daybreak to use it. The ring dial receives its importance from its small size (perhaps ten centimeters in diameter) and consequent portability.

¶ Optical technology was also making its impact in other areas during the Middle Ages. By the end of the thirteenth century, spectacles were in use. It is recorded that a certain Friar of Pisa preached a sermon in 1306 in which he claimed that he had spoken to the man who had invented, some twenty years previously, *eye glasses which make for good vision*. A portrait of the venerable Hugues de St. Cher wearing spectacles while seated at a desk copying a book was painted by Tommaso of Medina in 1352. This is believed to be the first portrait of a man wearing spectacles.

¶ The inventor of spectacles is shrouded in mystery, possibly because the inventor (if one existed) was supported by a duke or other noble and it would not have been prudent to bring the discovery to the attention of others and thereby invite competition and the wrath of the sponsor.

¶ Initially, only convex lenses were used in spectacles. These correct presbyopia (the inability to fully accommodate, from which many of us suffer with age). Reference to the use of concave lenses in spectacles to correct myopia (short sightedness) was not made prior to the mid-fifteenth century. In any event, the discovery of the utility of spectacles does not seem to have been accompanied by significant investigations of either the theory of lenses or of the functioning of vision.

¶ The optical devices described here were precursors of those that would provide data enabling our knowledge of the nature of light, and of the universe as well, to be impressively enlarged. Until the mid-sixteenth century, these devices aided man on his travels, his knowledge of the earth and solar system, and his ability to improve his vision. But it did so with limited precision. At that time, technology had improved sufficiently to enable optical science to make remarkable advances.

Precision prior to Telescopes

Tycho Brahe

THE NATURE OF LIGHT IS SO SUBTLE that a creative investigator must couple his ingenuity with instrumentation capable of achieving high precision to observe the minute nuances attributable to light. The ancients secured significant data with instruments which might be considered crude if judged by today's technology. From that data, however, they measured the size of the earth, erected precisely oriented monuments, deduced the height of the atmosphere, conjectured about the formation of rainbows, and studied image formation with lenses and mirrors. Nonetheless, these data were insufficient to permit a full description of what is meant by light. That required instrumentation capable of much greater measurement precision than that achieved by the sixteenth century.

¶ The epitome of precise measurements made without optical devices enhanced by lenses (such as the telescope) was obtained by Tycho Brahe (1546 to 1601). His measurements of stellar and planetary motions permitted a giant advance in our understanding of the universe, and led in turn to an improved understanding of light.

¶ Tycho was born in Denmark on December 14, 1546, as the second child and oldest son of an ancient and noble Danish family. Tycho's father promised his brother that he could raise Tycho (the first-born son) as his own, but he was reluctant to fulfill this promise when Tycho's uncle arrived to receive the child. Consequently, Tycho was carried off by stealth to be raised by his uncle. The parents accepted the situation, probably encouraged by the prospect that Tycho might inherit his uncle's wealth.

¶ Tycho was sent to the University of Copenhagen in 1559 to study rhetoric and philosophy in preparation for a career as a statesman. On August 21, 1560, Tycho saw a partial eclipse of the sun, and the event stimulated the young man (fourteen) by the precision with which men could foretell celestial positions. Thus was born a zest to study astronomy and science. However, Tycho's uncle admonished the boy not to stray from the studies that would lead toward the career chosen for him.

¶ After three years at Copenhagen, Tycho's uncle considered that the time had come for him to be sent to a foreign university. The University at Leipzig was chosen. To assure that Tycho would not digress from his prescribed studies and resort to scientific pursuits, a tutor accompanied young Tycho. Needless to say, the young scholar continued to pursue his chosen interest anyway. Moreover, a year after Tycho entered Leipzig, his uncle died. Despite the objections of his family, Tycho now pursued scientific goals openly.

¶ It is unnecessary to belabor Tycho's educational pursuits. However, it is interesting to note that a volatile temper and excessive pride had already developed in the young man—a characteristic that would follow him through life. During the Christmas season of 1566, Tycho and another Danish nobleman got into a quarrel at a party, renewed it at another, and finally resolved to settle their differences in a duel. As a result, Tycho lost a portion of his nose. To conceal his disfigurement, he fashioned a replacement with an alloy of gold and silver.

¶ By the age of thirty, Tycho was a renowned man. His astronomical observations were of great value, and he was also active as an astrologist and as an alchemist. At about this time, Frederick II, King of Denmark, recognized Tycho's brilliance and made a remarkable offer to him. He was given an island, Hven (called Venusia in Latin and Scarlatina by foreigners, Tycho says) and money to construct an observatory there. For twenty years, Uraniborg (Tycho's Temple of Astronomy) was a mecca for philosophers, statesmen, occasionally kings, and others from all over Europe.

¶ In the meantime, Tycho had married a peasant girl, much to the objection of his family. When there were eminent visitors, she sat at the head of the table. Tycho's pride and temper persisted. He was wont to contradict his guests, expose their ignorance, and snub them.

¶ But neither his ill temper nor his vanity diminished his brilliance as an observer. Tycho built quadrants, sextants, armillaries, parallactics, astrolabes, and other devices which he used to measure the position of the stars. Fortunately, he described these in his *Astronomiae Instauratae Mechanica*, published in 1598. It is appropriate to excerpt here from that work:

> We also had a very large quadrant made. It is called Mural, or Tychonicus . . . It is cast from solid brass and very finely polished. It is five inches wide and two inches thick, and the circumference is so large that it corresponds to a radius of nearly five cubits [194 centimeters]. Its degrees are in consequence extremely large and every single minute can be subdivided into six subdivisions; thus ten seconds of arc are plainly distinguishable and even half of this, or five seconds of arc, can be read without difficulty.

¶ Tycho continues in his description with reference to an illustration of the device. He notes that it contains a large portrait of himself "as if I were indicating to my collaborators what is to be observed." This portrait even includes a likeness of his dog resting at his master's feet.

¶ Tycho's arrogance and hauteur finally gave him more trouble than merely the loss of a nose. Tycho's lack of tact and diplomacy irritated many of his guests and neighbors. Consequently, when Frederick II died, there was strong agitation to force Tycho to leave. He left Denmark in 1597 and spent a brief period observing near Hamburg. He was then presented with the castle of Benatky near Prague, where he had his instruments shipped from Hven. In 1600 he was joined by Johannes Kepler (1571 to 1630), who was then not yet thirty. Their collaboration was short lived, for Tycho died within two years. However, he entrusted his work on his deathbed to Kepler, whose contributions are also monumental.

¶ Tycho's instruments and his methods of using them were certainly superior to any made prior to his time. With today's sophisticated technology, we may be a bit blasé about such precision. However, reflect on the fact that a man in New York subtends an angle of about one second to an observer in Boston. Reflect on the precision that you might attain—left to your own devices—and compare it with Tycho's achievements.

¶ In the hands of Kepler, the data amassed by Tycho permitted the determination of the earth's elliptic path about the sun. It also stimulated studies leading to the determination of the velocity of light. Many of the subtle characteristics of light which had been unobservable soon came under the scrutiny of investigators.

Tycho Brahe is pictured seated within the Tychonicus quadrant as if he were indicating to his collaborators what is to be observed. The quadrant had a radius of nearly two meters and permitted reading to an accuracy of better than five seconds of arc.

Glass Optical Instruments

The Phoenicians, Hans Lippershey, Jacob Adriaanzoon, Zacharius Jansen, George Huefnagel,
Francois Fontana, Anthony van Leeuwenhoek

 OST OPTICAL INSTRUMENTS IN USE TODAY employ glass components. How man learned to manufacture glass is conjectural. However, the origin of optical components made of glass is reasonably documented. Similarly, the impact such instruments made in scientific investigations is well known.

¶ Certainly, the ancients were aware of some of the properties of glass in refracting light. Pliny the Elder is said to have written that the Phoenicians were the first to manufacture glass on a large scale, having done so by about 3500 B.C. In his account, Pliny suggests that some Phoenician merchants discovered a method of manufacturing glass while they were preparing a meal on the bank of a river. Finding no suitable stones to support their saucepans over the fire, they used lumps of natron (an ore formed with native carbonate of soda). They were surprised to find that the natron melted from the heat of the fire and flowed onto the sands of the river bank. There it formed transparent streams of a liquid which, upon hardening, became glass. This story may be apocryphal, yet the Phoenicians did become famous for their glassware. Moreover, their skill and experience do not seem to have been surpassed during the next thousand years.

¶ To understand the apparent lethargy in the advance of glass manufacturing, one must consider two aspects. The first is the intricacy associated with glass manufacture. The second is associated with the demand for glass. Regarding the latter, glass was used primarily as an art form or as a vessel (such as a drinking glass) until optical instruments were conceived, beginning in about the fourteenth century. We shall discuss this shortly.

¶ Glass is a rigid but noncrystalline substance. It is made by fusing together alkalis (generally their salts), lime, and sand or flint. The fused mixture must be cooled slowly, becoming more viscous, until it congeals as a hard, clear solid. If the fused mixture is allowed to cool too rapidly, the silicates of the alkalis crystallize, resulting in an opaque, brittle solid. Hence, the process of glass manufacture requires a special oven in which the materials can be heated gradually to fuse and cooled slowly.

¶ Glass was manufactured in two stages. First, the materials were mixed in correct proportions and heated. In this fritting of the raw materials, the initial stages of the reaction eliminated some of the gaseous products and assisted the subsequent melting. In the second stage, the mixture was generally introduced into a second furnace where it was heated sufficiently to melt the mixture.

¶ Glasses have no sharp melting point. If the melting was not undertaken at a sufficiently high temperature, all of the gaseous bubbles did not escape and the resulting glass was either translucent or opaque. It is not difficult to see, therefore, why much of the early glass was unsuitable for optical purposes.

¶ By the later Middle Ages, glassmaking was undertaken in one of two somewhat distinct processes. In northern Europe, glass was made with local sands, which provided the silica, and with ashes of burnt inland vegetation to provide impure potassium carbonate for the flux. In southern Europe (primarily Italy) the silica was obtained by crushing white pebbles from the beds of rivers and the flux was a soda of impure sodium carbonate derived from burning marine vegetation.

¶ This division arose because of the isolation of the Rhineland and Gaul by the political upheavals that accompanied the collapse of the Roman power. The Gaulists were thus cut off from their former sources of soda and had to rely on potash.

¶ Certainly by the eleventh century, glass lenses were being made and used in scientific research. We have noted that Ibn al-Haitham and others studied refraction in glass and that by the fourteenth century, spectacles were in use. The well-established glassworks near Venice and in Holland provided the glass for spectacle makers in those localities and their trade flourished.

¶ Nevertheless, the invention of the telescope is still a matter of controversy. Claims are made for Hans Lippershey, an obscure Dutch spectacle maker. One story cites that two children (possibly Lippershey's) were playing with some lenses and chanced to note that by placing the lenses in a fortuitous position, the weather vane of a nearby church appeared to be larger than normal. Lippershey is said to have then duplicated this act and immediately realized the potential of the discovery. There are contradictory accounts regarding whether or not the lenses were placed in a tube, whether both lenses were convex or one was concave, and even how the idea originated.

¶ Lippershey petitioned the Dutch government in 1608 to be granted either an annual pension or other form of remuneration and a patent for his invention. He was commisioned to make a telescope, which he did, but his claim was placed under advisement.

¶ Meanwhile, Jacob Adriaanzoon, also of Holland, claimed in a similar petition that he had made a telescope like that of Lippershey and that, with proper encouragement, he could make a superior one. However, Adriaanzoon was apparently of an eccentric and jealous disposition and his claim may have been no more than a vain attempt to gain some form of recognition. He refused to show his telescope even to friends and caused his tools to be destroyed at his death so that his successors could not profit by his endeavors. His efforts were considered to be no more than rebuff.

¶ Another claimant to the invention was Zacharius Jansen, also of Holland. His son claimed that the invention took place in 1590, but others dated it as either 1611 or 1612. Jansen is reported to have been of a somewhat suspicious character. Besides being a spectacle maker, he coined false Spanish money in hopes of ruining that country's credit. He continued this practice after peace was declared, and was convicted of forgery. While waiting for the execution of the sentence, which directed that he was to be immersed in boiling oil, he fled the country.

¶ Because of the confusion regarding precedence, the patent rights which Lippershey claimed were refused. The news of the instrument spread quickly through Europe. By 1609 telescopes were available for sale in Paris, Frankfort, Milan, Padua, and London.

¶ It is not surprising to note that the microscope was invented essentially simultaneously with the telescope. William Boreel, the Dutch envoy at the Court of France,

wrote about efforts of Jansen in developing the telescope, noting that as a boy he was often in the Jansen shop, where he also learned of the invention of the microscope.
¶ As with the telescope, there are several claimants to the invention of the microscope. In addition to Jansen, George Huefnagel of Frankfort made such a claim and published several engravings of insects in 1592, possibly drawn with the aid of a microscope. Francois Fontana also claimed the honor in 1618, but this claim was certainly preceded by other claimants.
¶ The impact of the telescope on science was practically immediate. In contrast, the impact of the microscope was slow to be realized. Much of this impact can be attributed to the Dutchman, Anthony van Leeuwenhoek, and it is of interest to examine his career.
¶ Leeuwenhoek was born in Delft on October 24, 1632. In that same year, the philosophers Spinoza and Locke were also born, as was the architect Christopher Wren, and the painter Jan Vermeer. Anthony's father was a basketmaker, who died shortly after Anthony was born. Accordingly, Anthony was apprenticed to a linen draper in Amsterdam. He returned to Delft in 1654 to set up a linen shop and remained there for the remaining seventy years of his life.
¶ At some time, probably while still in Amsterdam, Leeuwenhoek became interested in microscopic observations and in 1673 he published a letter in the *Philosophical Transactions* regarding observations of the sting, mouth parts, and eye of the bee and the louse. These observations gained him considerable renown, and in 1680 he was elected as a Fellow in the Royal Society.
¶ Leeuwenhoek continued his interest in microscopic observations, and many scien-

This microscope, built by Anthony van Leeuwenhoek, measures about 4.5 by 2.5 cm and is capable of resolving details about 1.4 μm in width.

tists and other notables visited him to sight the microscopic world through his instruments. His microscopes consisted of brass plates riveted together. These measured about 4.5 by 2.5 centimeters. Between them a lens is fitted, having an effective diameter of but 0.5 millimeters. The object to be examined is fastened on the point of a needle fixed to one plate. In use today, such a microscope has been found to resolve details about 1.4 μm in width, which is close to the theoretical limit of 1 μm. This difference is probably due to scratches formed on the lens surface. Leeuwenhoek died on August 26, 1723, having achieved fame and association with many of the leading savants of the time. Moreover, he was responsible for bringing attention to the potential of the microscope.

¶ Optical instruments employing glass components are now commonplace. This was not so at the commencement of the seventeenth century, and the rapid advances made in understanding the nature of light during that century can be said to be attributable in large measure to the availability of glass optical components.　　　　　☺

The Dawn of Modern Science

Galileo Galilei

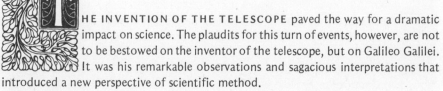

THE INVENTION OF THE TELESCOPE paved the way for a dramatic impact on science. The plaudits for this turn of events, however, are not to be bestowed on the inventor of the telescope, but on Galileo Galilei. It was his remarkable observations and sagacious interpretations that introduced a new perspective of scientific method.

¶ Galileo first heard of the telescope in May 1609. At that time he was forty-five years of age and a professor of mathematics at Padua. Already he had antagonized some of his colleagues with his outspoken anti-Aristotelian views. During a visit in nearby Venice, Galileo heard a rumor that an instrument had been constructed by a Dutchman that would enable one to see objects at a great distance as if they were near.

¶ Galileo immediately set out to construct his own telescope. By August, he had made one of sufficient quality to present it to the Venetian Republic. He was rewarded with a renewal of his contract—this time for life and at a substantial salary increase. Just how Galileo came to understand the principles involved with the telescope is not clear; to that time, he had not been occupied in optical research.

¶ Initially, the telescope was acclaimed for its military advantage. After all, it was demonstrated that a ship could be observed at sea several hours before it was evident entering a port. However, Galileo soon turned his telescope to the stars, observing that the moon had a rough, mountainous surface and that Jupiter had a round disk. By January, he noted three "stars" located in the vicinity of Jupiter. He at first assumed these to be in the background. A night later he noted that the position of these stars had changed and that they were joined by a fourth! Continued observations confirmed that these "stars" were satellites of Jupiter.

¶ Further observations of the moon confirmed that the surface of that body is, indeed, not smooth, as the Greeks had surmised. He also noted that the Milky Way consists of a multitude of stars, in fact of a "number . . . quite beyond determination." Galileo published the results of his observations in March 1610, producing an extraordinary sensation among the learned. It is of interest to note that Galileo did not interpret his observations in terms of the then-controversial Copernican theory. Nevertheless, the publication was met by strong opposition and adverse criticism; rabid Aristotelians denied his observations *a priori*. The misfortunes that followed are well known and need not be detailed here.

¶ Of interest in this anecdote is the character of Galileo himself. He was born on February 15, 1564, as the first child of a rather distinguished family. His forebears had taken part in the democratic government of Florence in the fourteenth century, and subsequent members of the family had distinguished themselves in medicine and law. However, the family's fortune had been largely dissipated by the time of Galileo's birth. His father was a musician and merchant, but a highly cultivated man of lively interests. He was well informed in the classical languages and in mathematics. Galileo seems to have inherited his father's independent character and combative spirit.

¶ Galileo's elementary education was humanistic. He spent some time in a monastery, presumably as a student. In September 1581 he entered the University of Pisa as a

This German etching depicts Galileo in prison in 1638, where he contemplated his teachings. Galileo was then seventy-four years of age.

student of medicine, which was his father's decision. Galileo displayed little interest in medicine and did not graduate as a physician.

¶ During the summer of 1583, Galileo was introduced to the study of mathematics, which excited the young man considerably. He soon began to undertake mathematical researches independently. Galileo considered mathematics as a potent instrument for acquiring a knowledge of nature. This naturally led to an interest in physics, which at that time consisted of Aristotelian teleological metaphysics. His passion for mathematics presented a conflicting influence in this pursuit.

¶ Very likely, his discovery in 1583 that the period of a pendulum is a constant did much to dissuade him from Aristotelian beliefs. This discovery came as a result of contemplating the swinging of a lamp in the Cathedral at Pisa, timing it with his pulse. As a result of this observation, he appears to have become imbued with the spirit of research by direct experiment.

¶ Even without a degree, Galileo was able to tutor mathematics in Florence. Yet, he longed for a teaching position with a university which would assure him of greater financial independence. He sought a position at several universities, finally obtaining the Chair of Mathematics at the University of Pisa in 1589. This was not an important chair and the pay was paltry, but it did offer Galileo an opportunity to display his abilities as a scholar.

¶ Galileo's conversion to Copernicism occurred in a rather unusual manner. While still at Pisa, three lectures were given on the Copernican doctrine by a follower of Copernicus. Galileo disdained attending these lectures. His fellow students, moreover, regarded the lectures as a subject of merriment, with one exception. Galileo considered this student to be both prudent and circumspect. Hence, he regretted that he had not attended the lectures, and he began to study Copernicus's teachings. Sometime between 1593 and 1597 Galileo became convinced of the truth of the Copernican system.

¶ Galileo's forthright attitude, coupled with his adherence to the Copernican doctrine, made many enemies for him. This, with the death of his father in 1591 (which placed upon him the responsibility for the family), compelled Galileo to seek a post at another university. Fortunately, he found a position at the University of Padua, although financially it was hardly more rewarding than his position at Pisa.

¶ With the telescopic observations already mentioned, Galileo's fame spread. At the end of March 1611 he traveled to Rome to exhibit his instruments to the ecclesiastical authorities. It was at this time, incidentally, that the term *telescope* was coined. This was accomplished by a man known as a poet rather than as a scientist. Galileo was received with the greatest honor by the Pope, Cardinals, and learned men. He was assured of the church's unalterable goodwill. Galileo returned to Venice early in June 1611, confident of his success.

¶ Politics and religion are topics that stir emotions frequently beyond intellectual comprehension. Galileo, as we know, found that out. Nevertheless, his contributions to science are practically unequalled. He inspired a new era of scientific investigation. Although the skeptic might argue that his direct contributions to optics were minimal, one cannot overlook the fact that Galileo's genius nurtured the seeds of subsequent investigations.   ☁

The Inception of Modern Optics

Rene Descartes

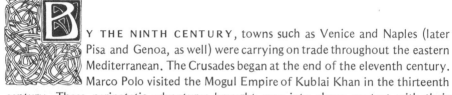 Y THE NINTH CENTURY, towns such as Venice and Naples (later Pisa and Genoa, as well) were carrying on trade throughout the eastern Mediterranean. The Crusades began at the end of the eleventh century. Marco Polo visited the Mogul Empire of Kublai Khan in the thirteenth century. These peripatetic adventures brought men into closer contact with their neighbors. At the same time, a renaissance of learning spread through Europe. Moreover, these contacts enabled Europe to become aware of Arabian discoveries and those Hellenistic achievements maintained in their repository.

¶ The study of optics was one of the dominant branches of science that prospered from the Renaissance. Some of this, it must be admitted, was based upon philosophical and metaphysical reasoning rather than on purely physical investigations. For instance, St. Augustine and other Neoplatonists regarded light as the illumination of the human intellect by divine truth.

¶ One of the first to apply scientific methods to the study of optics was Robert Grosseteste (1168 to 1253), although many of his views were also based in metaphysical concepts. As an example, he regarded light as the first form to be created from corporeal matter. Optics, according to this view, was *the* fundamental science. Grosseteste did couple this philosophy by advocating the experimental method.

¶ Another in this period to study optics was Witelo of Poland who experimentally determined new values for the angles of refraction of light passing between air, glass, and water. He probably noted that blue light is refracted more than the red, but if so, he did not follow up on this observation. He also contributed a discussion on the psychology of vision that indicated remarkable insight. However, much of his writing reflected the work of Ibn al-Haitham.

¶ Having greater originality than either of these two was Roger Bacon (1214 to 1292). He dissected vertebrate eyes and added knowledge to the understanding of optic nerves. He experimented with lenses to improve vision, but apparently with an imperfect understanding. At least he is attributed to claiming that Julius Caesar erected mirrors in Gaul to discern occurrences in England.

¶ Bacon believed that all branches of science are subordinate to theology, stating that knowledge consists of an inner sort which is of divine origin and a practical sort which is gained by observation and experiment. Together, he believed, they made up experience. His activity ranged throughout all aspects of science, prophesying mechanical transport on land and sea, aerial flight, and submarine exploration. In optics he made an attempt to revive Plato's concept of ocular beams, suggesting that something went forth from the eye when an object is viewed. Despite this, he stressed the importance of experiment, teaching that useful knowledge is gained only by elucidating the facts, rather than through unfounded speculation.

¶ One of the better-known figures in Renaissance science was Leonardo da Vinci (1452 to 1519). He was an illegitimate son and had no regular education. By some he is regarded as a disorderly and unsystematic thinker. However, his range of interest and his ability to observe made possible the numerous contributions associated with his life. He must have acquired some knowledge of the Greek philosophers and probably also read some of Ibn al-Haitham's work.

¶ Leonardo's interests were strongly artistic, which led into scientific studies. His anatomical observations were of an unsurpassed quality. Problems of perspective intrigued him, so he studied both geometrical and physical optics. He became acquainted with the structure of the eye and the function of its parts. He conceived of the eye as a form of *camera obscura*. Nevertheless, Leonardo's scientific achievements were minimal. His lack of scientific method robbed his ideas of any true fertility. He is remembered as a man of many interests and one of the keenest observers of nature of all times. His work provides us with insights into the great variety of problems under consideration by scientists at the end of the fifteenth century.

¶ In addition to these scientific achievements other important intellectual progress was being recorded as the Renaissance developed. Geoffrey Chaucer began the crystallization of the English language during the fourteenth century. The Black Death of the fourteenth century nearly depopulated Europe. In the aftermath, a life and death struggle arose between capital and labor, resulting in an elevation of the stature of the peasant. About 1450, Johann Gensfleisch Gutenberg began printing with moveable type, and books soon became widely available. In 1492, Christopher Columbus, confident of Ptolemy's calculations of the size of the terrestrial globe and desirous of finding another route to the Cathay which Marco Polo had described, set forth across the Atlantic Ocean. William Shakespeare was born in 1564, the same year as Galileo. Man was searching for his place in the scheme of things and his intellectual curiosity was whetted. Several gifted thinkers began to make a significant impact on the various branches of science that helped to bring European culture out of the doldrums of the Dark Ages.

¶ However, life at the close of the sixteenth century in Europe cannot be compared to what we today consider as civilized. The rulers of the day were governed by the ethics of the unconscienced, and the intellectuals and cultured were rarities. Religious bigotry and intolerance provided an incubator for further wars. The dispassionate pursuit of science was a hazardous enterprise.

¶ Born into this atmosphere on March 31, 1596, at La Haye, near Tours, France, was a boy destined to make important contributions to optics and mathematics. Rene Descartes was the son of Joachim Descartes, a Councilor in the Parliament of Rennes. His mother died at his birth. He was raised luxuriously and he inherited a fair fortune. Thus, Descartes grew to manhood somewhat oblivious to the political inclemency of the time. He was imbued with a modern spirit, longing to taste life to the fullest.

¶ In May 1617 he enlisted in the army of Prince Maurice of Nassau, the Prince of Orange. However, the campaign in the Netherlands ceased temporarily, leaving Descartes bored. He traveled to Frankfurt, where Ferdinand II was to be crowned. Enthused by the spectacle, Descartes again enlisted, this time under the Elector of Bavaria.

¶ While the army lay inactive in its winter quarters, Descartes found time for the tranquility and repose he sought. Then on the evening of November 10, 1619, Descartes had three dreams which he says changed his life. In the first dream, Descartes was blown by evil winds from the security of his college toward a third party which the wind was powerless to budge. In the second, he found himself observing a terrific storm with the objective eye of a scientist and noted that by doing so, the storm could do him no harm. In the third, he was reciting a poem which began *Quod vitae secatabor iter?* (Which way of life shall I follow?). Out of these dreams Descartes was filled with enthusiasm to pursue science that lasted a lifetime.

Descartes (1596 to 1650) during a walk in Amsterdam. *(The Bettmann Archive, Inc.)*

¶ With this new zeal, he came to a realization that truth was to be found only after first rejecting all ideas acquired from others and then relying upon the patient questioning of his mind. By the spring of 1621, Descartes had had his fill with soldiering (although he did serve once more in later years). He sought for a quiet life in northern Europe. However, while he was enroute by boat, the crew plotted to slay the wealthy passenger, loot his belongings, and feed his body to the fish. Descartes, however, understood their language and thwarted their plan by brandishing his sword and compelling the would-be assasins to return him to shore.

¶ Descartes arrived in Holland and spent a quiet year in developing his philosophy. We are familiar with his assurance of the thinking self, *cogito ergo sum* (I think, therefore I am). A year later he visited Italy, but failed to meet Galileo. Probably this can be attributed to Descartes's vanity. We can only conjecture on the contributions he might have made had he had the patience to exchange his views with the great philosopher.

¶ Descartes is best known for his pioneering work in analystical geometry—we still plot equations by the use of Cartesian coordinates. However, his contributions to optics were also of importance. He was the first to explain the nature of the rainbow, showing that at a particular angle the light is directed toward the viewer. He also studied image formation by lenses and showed that an aspheric surface would be devoid of spherical aberration, but the technology of the time could not produce such a lens.

¶ In these studies, Fresnel investigated the law of refraction. However, Willebrord Snell (1591 to 1626) had worked out the law in 1621. Snell, a mathematician at Leiden, appears to have been more interested in mathematical problems than in optical ones. He was engaged in determining the earth's radius by means of triangulation and may not have considered the derivation of the law of refraction of great value. Moreover, he derived the law in terms of ratios of length, interpretable as a ratio of cosecants. It is a matter of conjecture whether Descartes knew of Snell's work. Certainly, he was the first to express the law of refraction as a ratio of sines, having done so in 1637. Descartes based his conclusions regarding the law of refraction on mechanical analogies. He considered light to consist of particles which were accelerated along the normal to the surface when entering a denser medium. The resulting increase in velocity led to the use of sines. Sometime later, a fellow countryman of Snell, Christiaan Huygens (1629 to 1687), proposed a wave theory of light in which the velocity would decrease in the denser medium. This led to the same law, but the technology of the seventeenth century could not resolve whether the velocity of light increased or decreased in the denser medium. Huygens had access to Snell's notes and insisted that his countryman be given credit for having established the law of refraction. Apparently, there was some suspicion that Descartes had, indeed, plagiarized Snell's discovery and the law is known today as Snell's law. Descartes seemed to be uncertain about the nature of light, but he did make a statement which today seems somewhat prophetic. He suggested that light may be refracted by causes other than changes in the medium through which it was propagated, comparing this to the motion of a projectile. Did he thus anticipate Einstein by nearly 300 years in postulating that light rays would be bent when passing near a massive body such as the sun?

¶ In 1646, Descartes was living in happy seclusion in Holland. He meditated, gardened, and carried on an extensive correspondence with the intellectuals of Europe. And he was content to continue this existence. But the peaceful life was not to be. Queen Christine of Sweden had heard of him. The Queen, then nineteen, was a wiry athlete who took to the rigors of the Swedish climate with enthusiasm. She could stay on the saddle of a horse for ten hours without once getting off. She regarded others who could not maintain her pace with cold contempt. With this physical domination, she sought also to achieve intellectual greatness. So she invited Descartes to her court. Had he less of a streak of snobbery in him, Descartes would have refused. He not only longed for peace and tranquility, but his routine called for spending the morning in bed, where he wished to think. He recommended idleness as necessary to the production of good mental work. This was not what Descartes found in Sweden. Christine maintained that five A.M. was the proper time to study philosophy. At that time, in an icy library, Descartes began his lessons. That winter was regarded, even by the natives, as one of the coldest in memory. Descartes tried to alleviate the pain of his early lectures by resting in the afternoon. However, Christine decided to establish a Swedish Academy of Sciences which would meet in the afternoons.

¶ During the winter, Descartes fell ill of inflammation of the lungs. Doctors recommended that he be bled. Descartes at first resisted, but finally consented. It did no good. He died on February 11, 1650, at the age of fifty-four.

¶ We may consider that modern scientific enquiry dawned with Galileo, became inchoate with men such as Descartes, and bloomed in the seventeenth century.

The Solidification of Optics

Johannes Kepler, Francesco Maria Grimaldi, Ole Christensen Roemer, Robert Hooke

HY DID SCIENCE BEGIN TO BLOOM in the seventeenth century? There are many ways to answer this simple question. The intellectual reawakening begun with the Renaissance was broadening with a substantial impact on science. Shakespeare lived from 1564 to 1616. In France, Jean Racine (1639 to 1699) moved passions with his writings. John Milton (1608 to 1674) espoused the republican view in England. Francis Bacon (1561 to 1626) contributed to both philosophy and science, and attained political eminence.

¶ John Bernini (1598 to 1680) is celebrated for his skills as a painter, architect, sculptor, and mechanic. Fifteen of his pieces adorn the Church of St. Peter in Rome, including the altar and tabernacle, and St. Peter's Chair. Christopher Wren (1632 to 1723) was chosen as professor of astronomy in Gresham College in 1657. After the London fire of 1666, he proposed models for rebuilding the city. Although not entirely adopted, his efforts are evident in many landmarks, including St. Paul's Cathedral, Greenwich Hospital, Trinity College Library, and the theater at Oxford.

¶ Late in the sixteenth century, men such as Walter Raleigh (1552 to 1618) were leading settlements in America. Political upheavals were also evident in the seventeenth century. Oliver Cromwell (1599 to 1658) raised an insurrection which brought radical changes to the English nation.

¶ Perhaps as important to these events (as well as possibly contrbuting to them) was the spread of Protestantism which had begun with Martin Luther (1483 to 1546) in 1517. Man's regard for his destiny was certainly being re-examined.

¶ There were other subtle influences at work. Toward the end of the thirteenth century a postal system was established in France. By the sixteenth century, postal stations were instituted on the chief roads of France. In Germany, the first post was established in the latter half of the fifteenth century. In Italy, the transmission of letters was undertaken by individuals or communes prior to 1561. There were also private posts in England until 1635 when a public post was established. One might note that the first proper post in America was established in 1639 in Boston.

¶ Books were more freely available since Gutenberg. Thus, we see that information was able to flow with greater ease, there was more of it, and it was broadly available.

¶ With these brief reminders of historical events, it is in order to examine how our understanding of the nature of light was affected. The career of Johannes Kepler is a good case in point. He was born on December 21, 1571, in Weil, in the Duchy of Wirtemberg. Kepler's parents came from well-established families, but lost whatever wealth they possessed by poor management. Kepler's father was in the army and found it necessary to leave his pregnant wife to fulfill his military obligations. Johannes was born prematurely and was sickly in his early years. To join her husband, Kepler's mother left the infant with his grandfather. There followed periods of illness, family quarreling, occasions in which Kepler was obliged to perform the duties of a servant in his father's house, and finally his father's desertion and death. Accordingly, Kepler's early education can best be described as sporadic. Nevertheless, Kepler did manage to receive the degree of Bachelor in 1588 and the degree of Master in 1591.

¶ Kepler devoted his scholastic interests to philosophy generally, with an excellent background in mathematics, but none in astronomy. It might be considered odd, therefore, that he was appointed to the Chair of Astronomy at Graz, a protestant seminary, in what is now Austria, in 1594. There he set about trying to determine the number, the size, and motions of the planets. By this time, he had become a devout follower of the Copernican theory.

¶ He published his results in 1596 and they were brought to the attention of Tycho Brahe. As a result, correspondence was initiated and in 1600 Kepler visited Tycho in Prague. This resulted in Kepler's receiving an appointment as Tycho's assistant. Tycho died within a year, but Kepler was left with Tycho's exceptional observations. This enabled him to deduce his laws of elliptical orbits of the planets.

¶ While at Prague, Kepler undertook research on the geometrical properties of optics and the physiological processes associated with vision. He gave a good approximation to the laws of refraction and clearly described much of the physiology of the eye.

¶ Kepler also exchanged correspondence with Galileo, and in 1610 he obtained a Galilean telescope. He was able to study this instrument and carry his theory of refraction further to describe the image-forming capabilities of such a telescope.

¶ One of the major efforts which Kepler began with Tycho was the compilation of astronomical tables, to be called the Rudolphine tables after their benefactor. Although these were originally planned to be published in the first decade of the seventeenth century, they were not published until 1627. The immensity of the calculations makes it clear why the logarithms developed by John Napier (1550 to 1617) were so avidly received by Kepler. (Consider how he would have felt with a modern computer!)

¶ Thus we see that Kepler's career was enhanced by intercourse with other scientists. The intellectual awareness brought about by nontechnical cultural achievements also made its impact on science in the seventeenth century. Consider John Milton. He was raised by his father to undertake a life in the church. But Milton grew an aversion to the ecclesiastical profession. Instead, he devoted himself to the classics. His poetry brought him considerable fame. In 1638 he traveled extensively on the continent, meeting with persons of eminence, rank, and learning. Among these was Galileo. Milton's writings reflected his virulent attitude toward the church and his support of the republican principles of the time. One can but speculate how this was influenced by Galileo and the others, but it seems certain that some influence was exerted.

The first compound microscope was built by Robert Hooke. The instrument pictured here was photographed by the author at the Science Museum in London. It was formerly among the scientific instruments collected together by King George III. The gold stampings on the leather-covered body indicate that it was made by Christopher Cock, although it is very similar to the Hooke microscope described in 1665.

¶ It should not be thought that antagonism toward the establishment was the rule. Francesco Maria Grimaldi (1618 to 1663) entered the Society of Jesus at the age of fourteen and led the quiet life of a Jesuit scholar. In 1648 he was appointed to the Chair of Mathematics in Bologna. Grimaldi was a good experimental scientist. In one experiment he let sunlight into a darkened room through a narrow hole and placed another aperture in the path of the rays. The light patch on a sheet of white paper was found to be larger than the laws of rectilinear propagation would predict. In a similar experiment, he observed the shadows of thin opaque objects placed in the path of the rays. In doing so he found bright lines within the shadow. He attributed this to "diffraction." He attempted to define light in terms of a substantial entity with undulatory qualities.

¶ Another example of a scientist working within the establishment is that of Ole Christensen Roemer (1644 to 1710). He studied at the University of Copenhagen under Erasmus Bartholin (1625 to 1698) who had discovered the double refraction of light passing through a crystal of Iceland Spar. Roemer went to Paris in 1672 as a member of the French Academy. There he called attention to the fact that the period of revolution of one of Jupiter's moons determined by a series of immersions into the shadow of the planet differs from the period determined by the appearance of the moon from the shadow. Noting the difference in the distance of Jupiter from Earth during these determinations, Roemer explained the discrepancy on the assumption that light travels at a finite velocity.

¶ Roemer returned to Copenhagen in 1681 as a Professor of Mathematics. He soon became a member of the Supreme Court, the Director of the Police Forces, and Scientific Advisor to the King. In the last capacity, he reformed the Danish system of weights and measures, and introduced the Gregorian calendar.

¶ The increase in communication between scientists is also exemplified by the career of Robert Hooke (1635 to 1703). Hooke began his professional career as an assistant to Robert Boyle (1627 to 1691), who is well known for his experiments with the air pump. In 1662, Hooke was appointed Curator of the newly formed Royal Society, being responsible for the experiments carried out at its weekly meetings. In 1677 he became one of the secretaries of the Royal Society.

¶ Hooke hypothesized that light consists of an undulatory motion, even suggesting that the vibrations might be perpendicular to the direction of propagation. He invented the compound microscope. He made important contributions to astronomical instruments, being the first to advocate the importance of resolving power. He constructed the first reflecting telescope. In addition to these contributions to optics, he made numerous contributions to other branches of science. His interactions with scientists had direct benefit on their, and his, efforts. Moreover, he was instrumental in transforming the Royal Society from a group of scientific virtuosi to a professional body.

¶ Our anecdotes of events prior to the seventeenth century have generally considered a single individual. In the seventeenth century, isolated contributions became rarer. There began to be extensive interaction among learned men with their ideas furnishing provocation for another's discoveries. Perhaps that is why science began to bloom in the seventeenth century.

The Majesty of Newton

BY THE MIDDLE OF THE SEVENTEENTH CENTURY, observations of most of the manifestations of light had been made. The refractive properties had been relegated to a mathematical formula. The finite velocity had been measured. Diffraction and double refraction had been observed, although neither was understood.

¶ Telescopes and microscopes were being constructed with a variety of designs. Moreover, these designs could be based upon knowledge of the passage of light through such systems. The use of these instruments enabled man to discern that the Copernican system made much more sense than a geocentric system.

¶ A few hypotheses had been made regarding the nature of light, but they could only be regarded as speculative. The time was now ripe to replace suppositions with a firm theory that could explain the evidence uncovered by experiment. To enunciate such a hypothesis is perhaps beyond the ken of but one observer. Rather, it seems more probable that a genius must arise whose conclusions will instigate others to refine and polish his statements until a reasonable hypothesis is reached. At least, that is what happened.

¶ Isaac Newton was born on December 25, 1642 (the year that Galileo died), in the hamlet of Woolsthorpe. Newton's paternal ancestors exhibited no qualities that would indicate his genius; his father was known as "a wild, extravagant, and weak man." On the other hand, his mother was a woman of high character and fine intellect.

¶ At the age of sixteen, Newton was forced to take on the responsibility of managing the family estate. His father had died before he was born; his mother had remarried; his stepfather then died, leaving Newton's mother with three children by that alliance. Isaac was unequal to the task of yeoman farmer. Fortunately, his mother realized his unfitness for the task and sent him back to school to be tutored for college.

¶ Newton entered Cambridge in 1661, a period in which the school was struggling with disorganization from the upheavals resulting from the Civil War. Moreover, English schools, unlike their continental counterparts, had yet to respond to the impetus of the new knowledge derived from Galileo and Descartes. Even the influence of Francis Bacon had not shaken the grip of Aristotelian philosophy from the schools of divinity. Newton did not distinguish himself as an undergraduate, even failing a scholarship examination due to an inadequate preparation in geometry. During the academic year 1665/1666 he stayed in Woolsthorpe to avoid the plague. During this period he began experiments that brought him fame in optics, dynamics, and mathematics.

¶ He returned to Cambridge and by 1669 he was appointed to the Chair of Mathematics. He then began public lectures describing his observations while experimenting in optics. In 1672 he presented his first paper to the Royal Society and was elected a fellow that same year. That first paper dealt with his observations of optical phenomena.

¶ His first paper has been described as a work of art. It is a clear and concise arrangement of the pertinent results of his research, permitting the reader to grasp exactly what Newton had observed and leading the reader to a convincing conclusion. Newton did not stray by discussing observations of little relevance, although certainly some would have been made. It is to Newton's credit that he wrote so clearly; he had no example to follow.

¶ Newton's paper described how he had used a small hole in his "window shuts" to permit a small beam of light into his room. In the path of these rays he placed a prism

In this cottage Newton devised his theory of optics in the year 1665/1666 when he was forced to leave Cambridge to avoid the plague. The apple tree in the foreground is a scion of that which some consider to be responsible for the theory of gravitation. *(Photograph by the author in Woolsthorpe, Lincolnshire.)*

and observed the spectrum thus produced. The first thing he noted was that the colors so produced were in oblong form, rather than circular as one might expect from the law of refraction. He carefully experimented with the nature of the prism, the size of the aperture, and other factors to see if this apparent anomoly was introduced by some experimental cause. In doing so he even used a second prism to recombine the colors, noting that white light was restored. He considered the possibility that the rays might move in a circular path since, according to Descartes, the rays would consist of globular balls. Such would meet with greater resistance by the ether on one side and the trajectory would then bow as does a tennis ball when stroked with a spin.

¶ Newton concluded that dispersion would limit the perfection of refracting tele-

This statue of Newton holding a prism commemorates him at the University of Cambridge. *(Photograph by the author)*

scopes, which ultimately led him to adopt the reflecting telescope. He stated that as the rays of light differ in their refrangibility so they differ in their color. However, his paper did not conjecture on the nature of light itself. He limited his conclusions to "laws" describing the experimental evidence. Newton did not consider it proper to extend his experimental results to speculations on the mechanisms involved. That is, he would not pass from the field of physics to that of metaphysics.

¶ Although Newton's report was generally acclaimed, there were considerable opposition and criticism. After all, his findings contradicted some of the concepts of light then current. Hooke, for instance, complimented Newton on "the niceness and curiosity" of his observations, but disagreed with Newton's conclusions regarding the phenomena of colors. White light, from Hooke's viewpoint, is a "pulse or motion" and color is but a disturbance of that. Despite this divergence of viewpoints, Hooke and Newton conducted their disagreements in a gentlemanly fashion. Perhaps some of the bitterness that accompanied their exchange of views can be attributed to Henry Oldenburg. He was a German serving his country during Cromwell's usurption. Afterwards he became a member of the Royal Society, serving as assistant secretary with the responsibility for the publication of the *Transactions*. Oldenburg disliked Hooke and possibly used the disagreement between Hooke and Newton as an excuse to discredit Hooke.

¶ Another who voiced disagreement with Newton was Christiaan Huygens. He had been born in Holland in 1629 and was raised in an intellectual atmosphere that numbered Descartes among its visitors. Huygens had concluded that light is a vibratory motion in the ether spreading out from a source and inducing the sensation of light when received by the eye. He disagreed with Newton because the latter's observations were not explicable by his hypothesis of the vibrational nature of light. Ultimately, Newton and Huygens came to a mutual feeling of highest respect and regard for one another.

¶ There were other critics of Newton's observations. In one sense these seem to be trivial and perhaps motivated by jealousy. Many of these objections arose because Newton's experimental results contradicted a preconceived notion of the nature of light. The objectors refused to consider as factual any experimental evidence that would oppose their view. Newton responded to these objections by asserting that one must first "enquire diligently into the properties of things" and then "proceed more slowly to hypotheses for the explanation of them."

¶ One must not lightly dismiss these critics. They were far more considerate than the savage attacks on Galileo. Moreover, they were supporting the prevalent opinions of the day as advocated by such respected authorities as Kepler, Descartes, and Huygens. And Newton was yet an unknown.

¶ As Newton's genius became recognized, his followers became almost as avid as his critics had been. Newton's mechanistic philosophy (as manifested in his *Principia*) became evident in his *Optics*, published in 1709. Disputes now arose over the nature of light. Does it consist of corpuscles, or is it a form of wave motion?   ☉

The Dissemination of Optical Knowledge

James Ferguson

WHEN WE CONSIDER HOW THE UNDERSTANDING of the nature of light was being established prior to the mid-eighteenth century, we are likely to think only of the contributions of such distinguished individuals as Plato, Ibn al-Haitham, Galileo, Huygens, and Newton. Certainly, it was necessary to extend and verify their work, and numerous workers were needed to provide this, but the impact of their efforts generally was incidental to the mainstream of the development of optical knowledge. Accordingly, it hardly seems possible that an uneducated sheepherder could contribute to the erudition of optical knowledge in the mid-eighteenth century. But it happened.

¶ James Ferguson was such a person. He was born on April 25, 1710, a few miles from Keith in the north of Scotland (Keith is approximately forty miles northwest of Aberdeen). He was the second son of a day laborer who was barely able to provide for his family. Despite a sparcity of formal education, James Ferguson had a profound influence on the acceptance of the knowledge of optics through his lectures and writings. This influence was recognized in many ways, including his election to the Royal Society.

¶ Too poor to be sent to school, Ferguson learned his rudiments at home. To acquire a knowledge of reading, he surreptitiously studied his brother's Catechism. His father was pleased to learn of this endeavor and set about to teach the child to write, and then enrolled him in the grammar school in Keith. But the family's finances permitted only three months of this. Ferguson had no further formal education.

¶ His introduction to physics, which was to play an important role in his understanding of optics and astronomy, was received accidentally at the age of seven or eight. Part of the roof of the family house decayed and James's father set out to mend it. By applying a prop and a lever he was able to raise it by himself. Young Ferguson was astounded to see his father raise the ponderous roof. He attributed this at first to his father's strength, but soon realized that it was not strength alone that had been responsible for the feat, but rather the power of the lever. He extended this observation to studies of wheels and wedges.

¶ Pleased with his discoveries and desirous of sharing his knowledge, the boy wrote a short account of these machines, sketching out figures as necessary. He imagined it to be the first treatise on the subject that had ever been written. However, he discovered he was in error when a gentleman to whom he showed it let Ferguson examine a printed book that also discussed these principles. Nevertheless, Ferguson was encouraged to ascertain that his findings were in agreement with authority. Moreover, in this way he began a career, although he met several obstacles enroute that necessitated detours.

¶ Too young and too weak for hard labor, and with parents unable to provide for him, at the age of ten Ferguson was indentured to a neighbor to keep sheep. He then began studying the stars at night and constructing mechanical models by day.

¶ One evening, while lying in the field with a blanket drawn about him, he was observed to hold a thread at arm's length and to place beads attached to the thread so as to obscure from view certain stars. Then, laying the thread on a paper, he marked the relative positions of these stars, using a candle for illumination. His master, who had observed him, at first laughed at these antics, but upon learning of the meaning of the exercise, encouraged Ferguson to continue.

¶ Ferguson's next encounter with science occurred when he carried a message for his master to a minister in Keith. There he learned of the sphericity of the earth and examined maps for the first time. Upon visiting the minister one day, young Ferguson passed the schoolhouse where he had attended classes briefly and noted that a sundial was being painted on the wall. At the home of the minister, Ferguson was introduced to a neighbor, Thomas Grant. Grant was impressed with the lad and arranged that when his indenture was completed, Ferguson would live at the Grant residence.

¶ Ferguson, at the age of eighteen, completed his indenture and took up residence in the Grant household. Ferguson soon found that the butler, Mr. Cantley, was the man who had been painting the sundial on the schoolhouse wall. A friendship soon ripened and Cantley was able to teach his young friend much of value, for Cantley was a good mathematician; was well versed in music; understood Latin, Greek, and French; let blood well; and could even prescribe as a physician if needed.

¶ Two years later, Cantley left the Grants. Ferguson felt that he no longer had a reason to stay, so he returned to his home. There misfortune dogged the young man, resulting in a weakened condition of health. During his convalescence, he busied himself by making a globe, a clock, and other mechanical contrivances.

¶ The pendulum of fortune then swung the other way for Ferguson. Realizing that his father could not afford to keep him, Ferguson set out to find his fortune. He met some interested parties who admired his ability to draw patterns for needlework and to paint portraits. They took him to Edinburgh where he learned more about painting from the artists in residence. In addition, he found time to study anatomy, surgery, and other aspects of medicine by reading books.

¶ After two years in Edinburgh, the ever-confident Ferguson returned to his home feeling that he was now qualified to be a physician. But he soon found that he was not successful in this enterprise, so again he had to reset his course.

¶ In 1739, Ferguson obtained a job in Inverness that was more closely associated with his knowledge of astronomy. He set to work predicting eclipses, using his knowledge of the motions of the sun and moon. This effort was brought to the attention of the professor of mathematics at Edinburgh, who encouraged Ferguson to continue this activity. As a result, Ferguson moved again to Edinburgh where he was shown an orrery for the first time. But he was not allowed to examine the clock's works which were concealed in the wooden housing of the instrument. Nevertheless, Ferguson perceived how the gears should be arranged and he made such an instrument. As a result, he was invited to lecture to the students at Edinburgh and his career as a lecturer was launched.

¶ In the year 1743, Ferguson made another orrery of which all the wheels were made of ivory. In May of that year, he took it to London where he thought to further pursue his career as an astronomer. However, the position which he sought, and for which he apparently was qualified, was offered only to a bachelor. Ferguson had married four years prior to this, so was ineligible for the position. To support himself and his wife, Ferguson once again took up painting.

¶ But science was too ingrained in Ferguson to be cast aside. He was intrigued by the motions of the earth and the moon about the sun and soon made a simple machine to draw the relative path of the earth and the moon as a function of time. This was displayed on a long paper laid out on the floor. He carried the machine and tracing to Martin Folkes, at that time president of the Royal Society. Ferguson was asked to demonstrate this machine to the Society and he did so. One of the members invited Ferguson to visit him. There, Ferguson found that a machine such as the one he had demonstrated had been developed some twenty years earlier. The irrepressible Ferguson was not discouraged; rather, he found satisfaction that he had independently found the same solution.

¶ The following year, Ferguson published a pamphlet entitled *The Use of a New Orrery*, and this was quickly followed by several more publications. He soon became a popular lecturer and by the end of 1748 he abandoned his career as a portrait painter. He now pursued a career as a popular scientific lecturer and author. For his lectures, Ferguson constructed most of the apparatus he used to illustrate the principles he expounded.

¶ Ferguson soon authored several books, the most interesting (from the viewpoint of understanding the nature of light) was *Astronomy explained upon Sir Isaac Newton's Principles, and made easy for those who have not studied Mathematics*. This highly successful book presented, for the first time, a description of astronomical and optical phenomena in familiar language. Although Ferguson's scientific facts were not always correct, the book did much to stimulate the public on scientific matters. It was translated into Swedish and German, went through at least thirteen English editions, and was in demand for nearly seventy years. In addition to providing a stimulus to the general public, it helped to advance science, for it is said that the book did much to entice William Herschel to a career in astronomy.

¶ Ferguson contributed little that was original to optics, but he did much to spread the understanding of the nature of light. In line with the contemporary view, Ferguson adopted the corpuscular theory of light, stating that *Light consists of exceeding small particles of matter issuing from a luminous body*. He stated that the number of particles flowing from a candle in one second is 4.1866×10^{44}, and compared this to the number of grains of sand in the whole earth. He compared the velocity of light with that of a cannonball and concluded that if the bulk of the light particle was but a millionth that of a grain of sand, *we durst no more open our eyes to the light, than suffer sand to be shot point blank against them*.

¶ We can forgive Ferguson for the errors which crept into his book. Scientists would certainly ignore them and laymen would be impressed with the magnitudes he depicted. His merit as a teacher lay in the clearness of his prose, coupled with his ability to construct instruments and to prepare diagrams to illustrate his views. Moreover, Ferguson became a frequent visitor to King George III, where he discussed astronomy and optics.

Through this influence, support for the sciences was more generally forthcoming.

¶ Ferguson's life, as noted, was full of disappointments. Perhaps this prepared him for some of the misfortunes of his mature years. His family life turned out to be far from ideal. On one occasion, while he was away lecturing, his wife maliciously overturned several pieces of his scientific apparatus. His only daughter deserted him when she was eighteen, and her subsequent life is reported to have been disreputable. Ferguson's oldest son died at age twenty-four. Two younger sons were trained as surgeons; one never practiced, the other failed in his profession. Ferguson died in 1776, at age sixty-six.

¶ The search for an understanding of the nature of light has encompassed a variety of people. Some, like Newton, were men of exceptional genius. Some, like Huygens, were raised in an environment of science and in comfort. Some, like Tycho Brahe, had emotional problems with which to cope. Some, like Ibn al-Haitham, spent much of their lives incarcerated and their genius became manifest only in later years. James Ferguson had to overcome a primitive and isolated environment. His genius was shackled by misfortune, and never blossomed fully, for it was undernourished. Yet it is the interplay of the contributions each individual makes that brings fruition to any endeavor. Ferguson is to be admired for overcoming so many handicaps to provide substantial contributions to our effort to understand light.

The Achromatic Lens

PATENTS ARE A CURIOUS MATTER. Doubtless, invention prospers with the incentive proferred by a patent. Technology is certainly dependent upon invention and, therefore, in debt to the patent system. Since scientific advances often require improved technology to supply requisite data, it may be argued that science is also in debt to the patent system. Yet, the discoveries made in scientific areas are usually published in the open literature for all to share. In fact, the general character of a scientist's nature is to publicize his achievements. Seldom does a scientist hide his discovery in anticipation of personal gain. Exceptions exist, of course, and a consideration of the discovery of the achromatic lens might be cited as a case in point.

¶ The achromatic lens, which Newton had said could not be made, was perfected in 1733 by the barrister Chester Moor Hall. He practiced his art in secrecy and for nearly a quarter century prevented an invasion of his concealed capability. By 1758, John Dollond had found out the mystery, patented the process, and presented an account to the Royal Society.

¶ John Dollond was born in 1706 in London where his father had fled, along with many other Huguenots, after Henry IV of France had revoked the Edict of Nantes. John entered the weaving trade of which he became a master, but it is not as a weaver that we remember him. John had a strong interest in science which became manifested in his son, Peter. Peter was instructed by his father in mathematics and optics, rather than in weaving. With his father's financial backing, Peter opened a small optical workshop on Vine Street, Spitalfields (a district of London) in 1750. Two years later, John Dollond joined his son at The Sign of the Golden Spectacles in the Strand. We can imagine John Dolland's excitement as he left the rather humdrum affairs of weaving to begin optical experiments in the laboratory! And well it is for us, because the invention of the achromatic lens marked a tremendous advance in optical technology.

¶ Chromatic aberration was clearly described by Sir Isaac Newton in 1671. He showed that a lens does not refract light of different colors to a common focus. He concluded that it would be impossible to figure a lens to form a single image in heterochromatic light. Newton reached this conclusion because he erred in thinking that dispersion of the colors was in a fixed ratio to refraction.

¶ Chester Moor Hall, a barrister by profession, also dabbled in optics. By 1733, he discovered that dispersion and refraction are not in a fixed ratio. He was able to combine a concave component made of flint glass with a convex component formed with crown glass to form a single lens. This reduced the disturbance caused by the formation of differently colored images. Certainly, this was an invention of great consequence. Why didn't Hall secure a patent? One conjectures that he considered this to be but a novelty, and he was unaware of the potential this invention provided. Moreover, he was of a modest and retiring nature and demonstrated no inclination to publish his findings.

¶ Hall, however, did practice his art in secrecy. He ordered his crown and flint com-

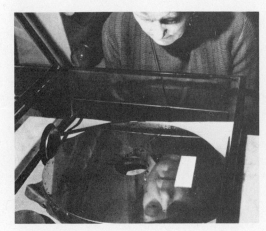

Reflecting telescopes of the 18th and 19th centuries were generally made of speculum metal (an alloy of copper and tin). As is seen in this photograph of Mrs. Lovell taken in the Science Museum, London, the reflectance was mediocre and the mirrors soon lost their figure.
(Photo by the author)

ponents from different opticians so they would not accidentally discover his technique. Nevertheless, as fate would have it, they in turn subcontracted the work, and quite coincidentally to one man, George Bass, a London glass grinder and polisher. The Dollonds also sent work to George Bass. During one of his visits, John Dollond noticed a piece of glass whose purity and luster interested him. Upon inquiry, Bass let it be known that this, together with another component, was to be used as a telescope objective. Dollond was quick to recognize the value of what he saw, and Hall's secret was out.

¶ Dollond set to work to investigate the relationship between refraction and dispersion of different glasses. He also applied for a patent, which of course was contested, but eventually was awarded to Dollond.

¶ In 1758, Dollond read a paper, *An Account of some Experiments concerning the different Refrangibility of Light*, to the Royal Society. Before this, in 1755, Samuel Klingenstierna of Uppsala had corresponded with Dollond regarding the achromatic lens. In 1760 he published the mathematical theory of the achromatic objective. John Dollond was honored as a Fellow of the Royal Society and received its highest award, the Copley Medal. As a result of Dollond's efforts, attention was soon paid by others to glassmaking technology. Improved telescopes were soon available. It is said that the Duke of Wellington spoke of the advantages Dollond's glasses gave him over the inferior French instruments. Of greater importance to science, observations were now possible that led to important advances in our understanding of the nature of light.

¶ John Dollond recognized both the scientific and technological importance of the achromatic lens. His scientific integrity prompted him to publish the results of his investigations. His perception of the value of this invention prompted him to secure patent rights. The patent protected him at a critical time so that he could firmly establish a flourishing concern. Still today the house of Dollond, now Dollond and Aitchison, is active in London. Historians will honor Chester Hall as the *inventor* of the achromatic lens. Scientists will honor John Dollond for *introducing* the achromatic lens.   ✸

A Seventeenth-Century Concept of Light

WHEN ONE ADVOCATES A VIEWPOINT, he generally proceeds by bringing to bear those aspects of the matter which will reinforce his position. In these anecdotes, the viewpoint being considered is the evolution in our understanding of the nature of light. It is, therefore, appropriate to consider how men discerned those properties of light that led to an understanding. It would be remiss of me, however, to suggest that observations followed in an orderly manner until the nature of light became evident. Indeed, there are many manifestations of light that tend to obscure rather than to illuminate (if you will pardon my selection of that word) our understanding of light. It is of interest to consider some of the observations that tended to obscure our understanding.

¶ Although published in 1780, the second edition of the *Encyclopedia Brittanica* (EB) reflects a seventeenth-century understanding of light. There, light is defined as *that invisible etherial matter which makes objects perceptible to our sense of seeing.* Following this, it is suggested that the nature of light can be described by either of two opposing views: the Cartesian or the Newtonian. According to the Cartesian view, *light is an invisible fluid present at all times and in all places, but which requires to be set in motion by an ignited or otherwise properly qualified body in order to make objects visible to us.* On the other hand, the Newtonian view considers light not to be a fluid *per se*, but to consist *of a vast number of exceedingly small particles shaken off in all directions from a luminous body.*

¶ The EB recognized shortcomings in both views. After all, if a luminous body emitted particles (the Newtonian view) it would, in time, vanish. On the other hand, if light consists of a wave (the Cartesian view) it would not continue to be propagated in a straight line upon reaching an obstacle *but will be always inflecting and diffusing itself every way to the quiescent medium beyond that obstacle.*

¶ Several manifestations of light are then described in the EB, (presumably) to permit the reader to grasp the nature of light. First, it was noted that most flowers follow the sun *by some power unknown to us.* The EB argues that it is the light rather than the heat which is the responsible agent for this, since flowers cultivated in a heated room turn toward sunlight rather than the fire used to heat the room. It was also noted that light is the responsible agent for the green color of plants. Moreover, they noted that solar light has the property of blackening precipitates of silver from nitrous acid. Thus, some of the fundamentals of the photographic process were known, but were not applied.

¶ The EB next discussed the emission of light independent of heat. Here, we encounter some of the complexities that tended to obscure an understanding of the nature of light. They reported light to be observed which was emitted from *putrescent substances and phosphorus*, citing: *At Montpelier, in 1641, a poor old woman had bought a pound of flesh in the market, intending to make use of it the day following. But happening not to be able to sleep well that night, and her bed and pantry being in the same room,* she observed so much light come from the flesh, as to illuminate all the place where it hung. A part of this luminous flesh was carried to Henry Bourbon, Duke of Conde, the governor of the place, who viewed it for several hours with the greatest astonishment.

¶ The EB took pains to show that it was not only "poor old women" who made such observations. Consider this account reported to have been experienced by Robert Boyle: *On the 15th of February 1662, one of his servants was greatly alarmed with the shining of some veal, which had been kept a few days, but had no bad smell, and was in a state very proper for use. The servant immediately made his master acquainted with this extraordinary appearance; and though he was then in bed, he ordered it to be immediately brought to him, and he examined it with the greatest attention. Suspecting that the state of the atmosphere had some share in the production of this phenomenon, he takes notice, after describing the appearance, that the wind was southwest and blustering, the moon was past its last quarter, and the mercury was at 29 3/16 inches.* Another observation of light emitted without heat reported in the EB of 1780 concerned a Father Bourzes who undertook a voyage to the Indies in 1704. There, he *took particular notice of the luminous appearance of the sea. The light was sometimes so great, that he could easily read the title of a book by it, though he was nine or ten feet from the surface of the water. Not only did the wake of the ship produce this light, but fishes also, in swimming, left so luminous a track behind them, that both their size and species might be distinguished by it.*

¶ Another manifestation of light without heat is also reported in the EB, which I find of particular interest. This is described as: *that luminous appearance which goes by the name ignis fatuus, or in common English Will with a wisp, to which the credulous vulgar ascribe very extraordinary and especially mischievous powers. This phenomenon is chiefly visible in damp places; and is also said to be very often seen in burying-grounds, and near dunghills. Travelers say, that it is very frequent near Bologna in Italy, and in several parts of Spain and Ethiopia. The form and size of it are very various, and often variable.*

¶ The EB reports several incidents of observing ignis fatuus, all quite mysterious and ghostlike. A similar report today would likely be disregarded by a serious scientist. After all, one of the objectives of science is to report observational results in a manner that will allow another to duplicate the observation. This is not to suggest that these observations were imagined; rather, it is to imply that it is difficult to interpret the observations in any scientific context because of the difficulty in controlling an experiment which will produce such manifestations of light. We still have sightings of unidentified flying objects (UFOs), but men today exhibit greater imagination by ascribing these to visitors from outer space rather than seeking a natural cause.

¶ At any rate, manifestations of the characteristics of light which would lead to a determination of its nature were sparse in the seventeenth century, and culling meaningful data from the unimportant is still a bewildering process. Is the light used to illuminate this page more *real* than the phantasmagoria of ignis fatuus or the perceptions received in a dream? Moreover, some of the observations of light phenomena made since the seventeenth century rival those discussed here in incredulity. Such observations are the basis of our present understanding of the nature of light.

"Hang on, Fred. It says here that it's nothing more than ignis fatuus." *(Cartoon by Fred Bleck)*

The Rise of Scientific Societies

THE EFFORTS OF THE PRE-SEVENTEENTH CENTURY scientists were largely individual undertakings. Intercourse with others was hampered by the scarcity of learned men and the slowness of the early postal systems. As scientific endeavors progressed, there was an increasing desire to discuss one's observations with one's peers. Gradually, men began to meet informally to discuss scientific achievements and to carry out experiments.

¶ Probably the first of these informal groups was an Italian society which arose in Naples, where Giambattista della Porta invited a number of learned people to his home, beginning in 1560. The group was limited to those who had made a discovery of a previously unknown fact in natural science. They met to discuss these achievements, which included a steam engine and a *camera obscura* but also consisted of a curious mixture of alchemy and magic. The meetings ended abruptly when della Porta was accused of witchcraft.

¶ Another of the early informal groups was the Accademia dei Lincei, which was formed in Rome by Duke Federigo Cesi in 1601. His interests lay in the study of bees and plants and he met regularly with three others to discuss these and other topics of mutual interest. However, they communicated in ciphers and this was met with suspicion by the authorities so the society was broken up. However, in 1609 it was reorganized on a larger scale and included in its membership both Galileo and della Porta. Since they planned "scientific, non-clerical monasteries" throughout the world for scientific cooperation, they excluded priests from membership. The plans included a museum, library, printing office, optical instruments, machinery, and laboratories.

¶ The proceedings of this society have the distinction of being the earliest (1609) recorded publication of scientific endeavors of any society. Two of Galileo's books were published by this society. Galileo made a microscope for use by the society and one of the members gave the instrument the name which is still used. This was certainly a glorious start for a scientific organization.

¶ However, the suppressive atmosphere of Aristotelian philosophy nurtured by the church soon provoked dissension among the members. A bitter quarrel arose regarding the acceptance of the Copernican doctrines. The condemnation of Galileo in 1633 made the study of physics and astronomy too full of dangers to be pursued further. Combined with the death of Duke Cesi in 1630, the society ended after less than a quarter century of existence.

¶ Literary societies also emerged in Italy during this period. However, the first organized scientific academy was the Accademia del Cimento of Florence. This was established by the Medici brothers, the Grand Duke Ferdinand II and Leopold, who supported it with both their wealth and enthusiastic participation.

¶ The Cimento had close ties with Galileo, although he had died before its inception in 1657. The Medicis had been pupils of Galileo as had been many of the early members. Moreover, the Cimento was dedicated toward experimental proofs and further elaboration of problems established by Galileo (and others). Nine scientists gathered at the Cimento to acquire experimental skills necessary for the determination of fundamental truths. Their efforts were welded together so that their work was sent into the world as if it had originated from a single individual. The book they published became, in essence, the "laboratory manual" of the eighteenth century, being the model and inspiration of other learned societies.

¶ In 1667, Ferdinand was given a cardinal's hat and the academy ended. There is a tradition that the Pope exacted its discontinuance for this award. With the extinction of the Cimento, Italy's leadership in physics ceased. The country was now reduced to supplying optical instruments, thermometers, and mechanical contrivances.

¶ In London, informal groups were meeting to discuss investigations in natural philosophy and other aspects of human learning by the mid-seventeenth century. One of these groups met weekly at various locations, including Gresham College. These meetings were interrupted during the insurrection when the building was converted to barracks for soldiers. But with the Restoration, in 1660, the meetings were revived. In November of that year it was proposed that consideration should be given to "founding a college for the promoting of Physico-mathematical Experimental Learning." Within a month they received notice of the king's approval and a formal agreement was signed by the original twelve members and seventy-three others, "to consult and debate concerning the promoting of experimental learning." On July 15, 1662, a royal charter was issued establishing the Royal Society.

¶ A pattern similar to that which occurred in England took place in France during this period. The famous literary Academie Francaise had been founded in 1635. Twenty-nine years later the Academie des Sciences was founded.

¶ The French society bore more resemblance to the Royal Society than to the Cimento, but there were several distinguishing features worthy of note. First of all, the English society, although granted "Royal favor, patronage, and all due encouragement," was given no financial provision whatsoever. As a result, for many years the Royal Society faced pecuniary struggles that threatened its existence. The Academie des Sciences, on the other hand, had the advantage of the resources of the royal treasury. Members of the French society thus drew fixed pensions, means were supplied for laboratories and instrumentation, and foreign scholars were attracted to Paris. It is small wonder that French science prospered so well.

¶ The work performed by the two societies also differed. In France, the experiments were discussed in advance, then jointly performed in the laboratory. The British scientists selected their projects individually and worked in their homes. Perhaps the cooperative effort is not the optimum method of achieving significant results. The discussions that took place in Paris sometimes dragged on for weeks due to the "talkativeness" of some members. Jealousies also developed which hampered collective research.

¶ The Royal Society attracted the highly qualified savants, many public dignitaries (noblemen, knights, bishops, and doctors of divinity), and amateurs. Indeed, particular appeal was made to the latter since they "bring not much knowledge, yet bring their hands and eyes incorrupted." Hooke emphasized the advantage of bringing merchants and businessmen into the ranks to gain thereby financial help.

A

TREATISE

OF

OPTICS.

BOOK I.

The Elementary Part of Optics.

THE object or defign of OPTICS is to explain the *Theory of Vifion,* as it is founded upon known and eftablifhed properties of light. Thefe properties may be confidered in the abftract, that is, without an immediate application to Vifion; and this properly makes the elementary part of the fcience: for upon thefe are founded the furprifing effects of all forts of optical inftruments; the principal of which are *Microfcopes* and *Telefcopes.* Both thefe enlarge the apparent magnitudes of objects: the Microfcope being adapted for magnifying fmall objects that are near at hand; whilft the Telefcope is applied to remote objects.

Optics is diftinguifhed into two principal heads: *Catoptrics,* which treats of the reflection of light by fmooth or polifh'd furfaces of given figures; and *Dioptrics,* whofe object is the progrefs of light through tranfparent bodies, terminated alfo by given furfaces. But of all the kinds of furfaces, the moft ufeful in Optics are the *Plain* and *Spherical,* and to confider others is feldom neceffary: nor is it requifite to make a formal diftinction as to Catoptrics and Dioptrics, it being often more advantageous to blend them together.

SECTION I.

Of the chief Properties of Light, with an Explanation, as they occur, of fome general Terms.

Art. I. IT is manifeft that rays of light are inceffantly propagated in all poffible directions, from every phyfical point of a luminous body; and when they enlighten other bodies upon which they fall,

B

they

The first page of A TREATISE OF OPTICS by Joseph Harris. This book was sold by B. White in London in 1775. *(From the author's library)*

¶ Of interest to us are some of the fellows added to the society. These include Huygens (1663), Newton (1671), Flamsteed (1676), Halley (1678), and van Leeuwenhoek (1679). Flamsteed (1646 to 1719) was a self-taught astronomer who became the first Astronomer Royal of the Observatory of Greenwich in 1675. The original cause of the alliance of these organizations lay not in their common interest but in poverty and the need to borrow instruments. Halley (1656 to 1742) is known for his calculation of the orbit of Halley's comet.

¶ Newton was elected on January 11, 1672, and soon communicated to the Society his invention of a reflecting telescope. Newton's next contribution concerned his theory of light and colors. It was through publications of the Royal Society that these discoveries were made known outside of Cambridge. It is safe to say that the scientific career of Newton was materially influenced by the Royal Society and that the converse is also true.

¶ The French scientific discoveries were also broad and influential. Huygens was invited to Paris soon after the establishment of the Academie. Roemer was also brought there, where he computed the velocity of light from observations of the occultation of the moons of Jupiter. However, in 1683 the Academie entered a period of decline during which academicians were degraded to serving the curiosities of the king and state. This period lasted for just over a decade, at which time the Academie received a new constitution that encouraged scientific progress. The Academie was then fruitful until it was dissolved during the Revolution in 1793. One might compare the fortunes of the Academie with the oscillations in intellectual progress that occur in this country as political values oscillate.

¶ In retrospect, we have seen that scientific societies seem to have originated in the various countries at about the same time. Their evolution, however, was quite dissimilar. The Cimento originated at the behest of the Medici; the English and French societies sprang from informal and spontaneous gatherings. The Cimento was directed specifically toward scientific study; the others were broader in scope, incorporating considerations of trade and commerce and manufacture. The Cimento featured a corporate effort; the English and eventually the French directed their efforts toward personal ambitions. The English and French societies are also distinguished by their form of organization, which became the model of the Berlin Academy, founded late in the seventeenth century, and of many learned societies founded subsequently.

¶ About the only moral that I can draw from the different ways in which the societies grew is that men profit from alliance, but fortunately we are motivated by different ideals. *Vive la difference!*   ☙

The Discovery of Invisible Light

Friedrich Wilhelm Herschel

SCIENCE, BY THE END OF THE EIGHTEENTH CENTURY, was making remarkable progress. Nevertheless, as investigations opened one door to provide a glimpse of physical reality, other doors were exposed to present a conflicting interpretation of the significance of the vista pieced together from this complexity of views. Certainly, this was true of the search to understand the nature of light. While Plato's ocular beams were no longer considered a reasonable hypothesis, light was considered by some to be the transmission of small particles—of corpuscles—while others advocated some sort of wave motion. In either event, light was responsible for vision and thus was correlated with this sensory response. One could hardly conceive at the end of the eighteenth century of *invisible light*. But in the year 1800 such was reported, leading to a new comprehension of light itself.

¶ Sir William Herschel (1738 to 1822) was the first to report having isolated and detected radiation beyond that sensible to human vision. He did this in 1800 in a well-planned and carefully executed experiment. However, the significance of light which was not evidenced by the eye could not be readily accepted.

¶ Friedrich Wilhelm Herschel was born in Hanover on November 15, 1738. His father was a musician, his paternal grandfather a master gardener, and his great grandfather a brewer. Thus, his environment was respectable, but not prosperous. His father supplied him with little more than the bare domestic necessities, but he did do his utmost to stimulate Herschel's mind with an appreciation of the splendors of nature.

¶ Herschel had what must be termed a sparse and primitive technical education, but he was influenced to adapt his alert and receptive intelligence to cultivate knowledge for its own sake. He followed his father's footsteps into the Hanoverian military band. In 1757 Herschel was exposed to his first experience of battle, which lost for him any taste he might have had for the regimental band. Although not a soldier, Herschel quit his unit, quit the continent, and sought refuge in England.

¶ Herschel soon settled down to a variety of musical assignments. As much as Herschel seemed to enjoy life as a musician, he apparently preferred astronomical observing, which began in earnest at this time. His interests soon migrated to the stars rather than to the planets of the solar system. This required telescopes of larger aperture to gather the feeble light received on earth. Large apertures meant reflectors rather than refractors since large pieces of glass could be neither poured nor polished. Accordingly, Herschel set out to fashion his own mirrors of speculum metal.

¶ On the night of March 13, 1781, Herschel was observing the sky in search for double stars from which he might determine parallax and astronomical distance. While doing this, he came across an object that was clearly (to him) unlike a star. He suspected that it was a comet, but additional observations showed this to be a theretofore unobserved planet, subsequently named Uranus.

¶ An account of this discovery was read to the Royal Society, and it excited great interest. Scientists of the day had been lulled into thinking that the important scientific observations began with the publication of Copernicus's *De Revolutionibus* in 1543 and had been completed when Newton published his *Principia* in 1687. Herschel's observation of a new planet doubled the extent of the known solar system and gave evidence that much unknown remained to be solved.

¶ As a result of this discovery, Herschel was elected to membership in the Royal Society, awarded the Copley prize, and appointed as private astronomer to King George III. Herschel now had time to devote to his intensive surveys of the sky. In 1784 he began his description of the construction of the heavens. His reputation by this time was extensive, but with restless energy he persisted in his observations.

¶ In 1800 Herschel (at the age of sixty-one) noted, while making telescopic observations of the sun, that various colored glasses used as filters over his eyepiece gave quite different relief from the heat to which his eyeball was subjected. He remarked, in a paper to the Royal Society, that *when I used some of them, I felt a sensation of heat, though I had but little light; while others gave me much light, with scarce any sensation of heat.* He, accordingly, began a simple experiment to determine how the heating power and light are distributed in the solar spectrum. His procedure was straightforward: he placed a prism in a window where it intercepted sunlight and the refracted rays were received through a rectangular opening of size sufficient to pass one color.

¶ For a detector, Herschel selected one of three thermometers, the bulbs of which he had initially blackened with Japan ink. The remaining two thermometers were placed close to this detector to ascertain if the ambient was disturbed. He noted the rise indicated by the selected thermometer when exposed successively to the various colors formed by dispersing the sunlight. Herschel, of course, had no way to determine the nonlinear characteristics of the prismatic dispersion and thus came to the natural conclusion that *the heating power of prismatic colors, is very far from being equally divided, and . . . the red rays are chiefly eminent in that respect.*

¶ He next undertook to determine the illuminating power of the various colors. He considered this power to be equal to the ability of each color to provide a good image. His experimental method, again quite straightforward, consisted of examining with a microscope various objects illuminated by a single color of dispersed sunlight. From repeated examinations of several objects with the microscope, he concluded:

> . . . *with respect to the illuminating power assigned to each colour, we may conclude, that the red rays are very far from having it in any eminent degree. The orange possess more of it than the red; and the yellow rays illuminate objects still more perfectly. The maximum of illumination lies in the brightest yellow or palest green. The green itself is nearly equally bright with the yellow; but, from the full deep green, the illuminating power decreases very sensibly. That of the blue is nearly upon a par with that of the red; the indigo has much less than the blue; and the violet is very deficient.*

¶ Herschel had thus presumably accomplished his mission. His results would suggest, for instance, that green light is capable of good illumination, but is incapable of strong heating. Therefore, a green (or blue) filter would be indicated to serve his purpose in

William Herschel measured the heat in the solar spectrum with three thermometers, two of which were not exposed to the light but were used to check any deviations in the ambient temperature. *Plate XI of Phil. Trans. 1800, p. 292.*

seeing the sun without unduly heating his eye. Indeed, he devoted ten pages of his paper to a description of tests he made with various filters to support his findings.

¶ It would be unfair to imply that Herschel was not guided by a reasonably thorough understanding of the significance of his observations—at least "reasonably thorough" within the state of understanding of the time. However, several, whom we would consider to be competent authorities, considered that radiation existing beyond that sensible to vision was an absurd notion. Herschel gave no hint that he considered the detection of radiation in the spectral region we now call infrared to be at all remarkable. In fact, he failed to describe his initial observation of heating effects beyond the red. Only in summarizing the results of his observations did he comment that *I likewise conclude that the full red falls still short of the maximum of heat; which perhaps lies even a little beyond visible refraction.*

¶ Herschel, at that time, seemed to be on the verge of an explanation acceptable to modern concepts, writing:

> *In this case, radiant heat will at least partly, if not chiefly, if I may be permitted the expression, of invisible light; that is to say, of rays coming from the sun, that have such a momentum as to be unfit for vision. And, admitting, as is highly probable, that the organs of sight are only adapted to receive impressions from particles of a certain momentum, it explains why the maximum of illumination should be in the middle of the refrangible rays; as those which have greater or less momenta, are likely to become equally unfit for impressions of sight. Whereas, in radiant heat, there may be no such limitation to the momentum of its particles.*

¶ Herschel's genius is manifest both in having made this remarkable observation and in his quickly perceiving rather clearly the nature of the phenomena that he had confronted. Nevertheless, accepting the fact that light could be invisible was revolutionary. Herschel displayed a hesitancy later in adhering to his original conclusion. Others attacked him vigorously, and somewhat unmercifully, but subsequent independent observations confirmed Herschel's original conclusion regarding the validity of invisible light.   ☢

Does Invisible Light Exist?

THE DISCOVERY OF "INVISIBLE LIGHT" by William Herschel in 1800 not only incited contradictory responses from his peers, but caused Herschel to doubt its significance. Two decades earlier, the tenor of the times was such that his discovery of a new planet made considerable impact. But now scientists had become accustomed to the revelations of additional discoveries. The concepts of science were undergoing frequent revision. Nevertheless, invisible light seemed a contradiction of terms.

¶ To appreciate the reaction to Herschel's discovery, one must attempt to understand the accepted theories of light and of heat of that time. Herschel had gained his knowledge of optics primarily from studying *A Compleat System of Opticks* by Robert Smith. Smith advocated the Newtonian concept of corpuscles, and considered light to consist of *very small and distinct particles of matter* [which] *impress upon our organs of seeing that peculiar motion, which is requisite to excite in our minds the sensation of light*. He considered sunlight to be *a mixture of several sorts of coloured rays, some of which at equal angles of incidence are more refracted than others, and therefore are called more refrangible*. None of this prepares one for invisible radiation. In fact, just the opposite conclusion would be drawn.

¶ Others had dispersed sunlight and determined the heating characteristics in the resulting spectrum, although Herschel made no reference to their work. Naturally, the extent of the spectral sensitivity of the eye had not been determined in 1800. In fact, little more regarding visual perception was then known than had been advanced by Newton nearly a century and a half earlier. Smith noted that *the anatomist when they have taken off from the bottome of the eye that outward and thickest coat called the dura mater, can see through the inner coats the pictures of objects lively painted thereon. And these pictures propagated by motion along the fibres of the optick nerves into the brain, are the cause of vision*. This statement was in full agreement with Newton. Similarly, Smith was unable to improve on Newton's concept of color vision, agreeing that colors make vibrations of "several bignesses."

¶ A similar text entitled *A Treatise of Optics* written by Joseph Harris and published in 1775 compares sight to the other senses. He quoted Newton as saying, *Do not the rays of light falling upon the bottom of the eye excite vibrations in the tunica retina? Which vibrations, being propagated along the solid fibres of the optic nerves into the brain, cause the sense of seeing*. Harris then agrees with Newton that since dense bodies conserve their heat the longest, the vibrations caused may be propagated a longer distance to the brain. Accordingly, heat is conveyed by our sense of touch, but presumably the fibers connecting the retina to the brain would not be dense enough to convey heat.

¶ Herschel's writings suggest that he must have been aware of the contradictory nature of his discovery. Immediately upon concluding his observation of radiant heat in the refracted solar spectrum and reporting this to the Royal Society, he began additional experiments. There was no doubt then that he was convinced of the validity of invisible rays, for he wrote:

> *It remains only for us to admit, that such of the rays of the sun as have the refrangibility of those which are contained in the prismatic spectrum, by the construction of the organs of sight, are admitted, under the appearance of light and colors; and that the rest, being stopped in the coats and humours of the eye, act upon them, as they are known to do upon all other parts of the body, by occasioning a sensation of heat.*

¶ Herschel set out to investigate the similarities of light and heat. His actions reflected the Newtonian scientific method. The preliminary observations suggested tentative hypotheses to describe the nature of thermal radiation. It would be in order to test this by experiment, comparing the results with the known characteristics of light.

¶ He began by observing that both light and heat obey the laws of reflection. He did this by noting the reflection of heat of the sun in a Newtonian telescope, the reflection of the heat of a candle flame using a small concave steel mirror, and similar experiments. These were convincing demonstrations that both light and heat obey the laws of reflection.

¶ He next undertook a series of experiments to indicate if both light and heat are subject to similar laws of refraction. Following this, he began experiments to indicate that heat is subject to various refrangibilities as is light. But with light one can observe that solar light is dispersed into colors. How can one differentiate the components of dispersed thermal radiation?

¶ Herschel had apparently considered that these experiments would be completed quickly. But after submitting three papers to the Royal Society in rapid succession during the spring of 1800, his fourth paper was not submitted until November of that year. There was now growing doubt evident in Herschel's communication. He stated, *we should at the same time point out some striking and substantial differences* [between heat and light].

¶ To illustrate one of these differences, Herschel drew a solar spectrum using as an abscissa the colors of his dispersed spectrum and as an ordinate his subjective impression of the illuminating power of these colors. He then extended this spectrum by a distance proportional to the distances he had placed his thermometers in detecting the heat of the solar spectrum. Herschel was thus able to superimpose a second spectrum over the first, this one incorporating the temperature rise observed as the ordinate. The two spectra are markedly different.

¶ Herschel attempted to apply Snell's law to the thermal radiation, but this effort is clearly meaningless in the spectral region beyond the red end of the visible spectrum. His next series of experiments consisted of determining the transmissive and absorptive characteristics of various materials to light and heat. To measure the reflection of a sample, Herschel used a photometer previously described by Pierre Bouguer (1698 to 1758), who had laid down the fundamental laws of photometry in 1729. These principles had been further elaborated by Johann Lambert (1728 to 1777) in 1760. However, Herschel used this apparatus to compare the brightnesses of different hues, which is at odds with the theory. One would suspect, in any case, that his instrument

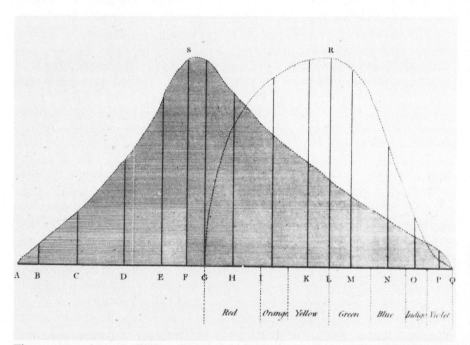

These spectra, drawn by William Herschel, indicate the heating and illuminating power of the solar spectrum. The spectrum R was prepared by Herschel's observations of the illuminating power. The spectrum S was prepared by observing the temperature rise of a thermometer placed at that point in the dispersed radiation. *(From Phil. Trans.)*

was incapable of high accuracy. Nevertheless, Herschel presented his data on the assumption that 1000 rays were incident on a sample, thus implying a precision of 0.1 percent or better.

¶ The samples Herschel used in these experiments were objects readily available in a laboratory. There were several glasses, many of which were colored. He also used a tube mounted over one of the thermometers as an absorption cell to measure the transmission of available liquids. Here, Herschel found that his gin had a transmission less than 50 percent of light and less than 30 percent of heat. On the other hand, his brandy was nearly opaque, but transmitted about 20 percent of the heat.

¶ Unfortunately, Herschel's poor data (largely derived from a poor experimental technique), coupled with the then-prevalent (and incomplete) concept of vision, led him to conclude that light and heat *have nothing in common but a certain equal degree of refrangibility*.

¶ Nevertheless, Joseph Banks (1743 to 1820), president of the Royal Society, wrote to Herschel on March 24, 1800, *We are all struck with the discovery of Radiant Heat being seperable from Radiant Light*. Two weeks later he again wrote, *I consider the separation of heat from light as a discovery pregnant with more additions to science* [than the discovery of a new planet].

¶ John Leslie (1766 to 1832), who had been experimenting with radiant heat in Scotland, viewed Herschel's discovery as *not consistent with strict metaphysics; and . . . surely stretching the limits of probability*. Leslie suggested that Herschel was guilty of fallacious observations, stating, *a more objectionable plan of conducting experiments could scarcely have been devised*. Leslie was commenting here on Herschel's placement of the thermometers. Leslie's continued criticisms of Herschel became so abusive and somewhat pointless that as one reads them today, they seem comical. One suspects that he was persuaded by some sort of religious compassion. But perhaps Leslie's criticisms merely reflect the scientific uncertainty of the time. Herschel had been erratic in his interpretation of the data. This is a subject always open to a difference of opinion.

¶ Leslie's challenge of the validity of Herschel's experiment could readily be resolved. Henry C. Englefield (1752 to 1822) was asked to repeat Herschel's experiment at the Royal Institution. He began his experiments in the spring of 1801 before two witnesses. He first verified Herschel's assertion that the maximum heating is observed beyond the red end of the spectrum. He next dispersed sunlight onto the hand of one of the witnesses, who had been blindfolded. When asked to locate the maximum heat, the witness always touched his hand beyond the red extent of the dispersed radiation. These experiments were repeated in 1802 for Humphrey Davy (1778 to 1829). Herschel was thus vindicated.

¶ The concept of invisible radiation was further enhanced in 1801 by Johann W. Ritter (1776 to 1810) who discovered chemical actions incited by the action of dispersed radiation beyond the violet end of the visible spectrum. Ritter wrote, *These enquiries form a sequel to the experiments by which Herschell* [sic] *discovered the existence of invisible calorific rays beyond the limits of the solar spectrum.*   ☙

Science Begins to Mature

N THE SEVENTEENTH CENTURY, all branches of science, including optics, showed a remarkable development. It was largely a period in which science was placed on new foundations as the sterile tradition under which science had languished was destroyed. From the publication of Gilbert's *De magnete* in 1600 to Newton's *Principia* in 1687, the face of science changed almost beyond recognition.

¶ Scientific societies were formed, many with support from the government or with aristocratic patronage. Newton's "experimental physics" now spread from England and Holland to the rest of Europe, resulting in spectacular advances in the study of electricity, magnetism, chemistry, and geology—as well as mechanics and optics.

¶ The spirit of the eighteenth century was one of scientific enquiry. The enlightenment shifted men from seeking revealed truths to reason. Trade routes also changed. Rather than looking to Asia for exotic products, the maritime powers were drawn to the Americas. Accordingly, the center of European culture shifted from the Mediterranean to the shores of the Atlantic. The eighteenth-century man also had a change in values. Manners were easier and intellectual adventure was the order of the day.

¶ The scientific scene of the eighteenth century was dominated by Newton's brilliant explanation of the doctrines of Copernicus and Kepler. The importance of this extended beyond astronomy and physics. Natural philosophy was inspired with a new vision in which the order of the universe could be determined by a set of equations.

¶ The expanding needs of daily life turned craftsmen into inventors, thus adding further impetus to the vision. Street lighting was introduced, pavements were provided for pedestrians, houses became designed for greater comfort. By the end of the century, many houses had running water. In addition, abstract scientific studies led to improvements in practical life. Astronomical investigations led to improved methods of determining longitude at sea. Musical instruments were improved as a result of a better knowledge of mathematical acoustics. The foundations of the industrial revolution were being laid.

¶ To describe all of the scientific activity that took place in the eighteenth century would require an additional compilation of anecdotes. Here, we must be content to skim the highlights so that optical advances can be placed in clearer perspective.

¶ From the earliest times, it was known that when a "hot" and a "cold" body are brought into proximity, the former cools and the latter warms until they reach a common temperature. Moreover, it was known that most bodies expand when heated. It was not until the seventeenth century, however, that Galileo and others used thermal expansion in the construction of thermometers. Gabriel Daniel Fahrenheit (1686 to 1736) brought the technology of making thermometers into general use by introducing mercury as the expanding substance rather than a gas. Fahrenheit assumed blood heat to be 100 degrees and the coldest available freezing mixture to be 0 degrees. On this scale, the freezing point and the boiling point of water become 32 degrees and 212 degrees.

When the latter value was accepted as a standard, it was determined that blood heat is not quite so high as Fahrenheit had thought.

¶ Rene de Reaumur (1683 to 1757) was an active member of the Paris Academy of Sciences for fifty years. During that period he contributed broadly to the application of science to industry. In particular, his researches were responsible for the establishment of steelmaking in France. However, he is perhaps best remembered for his thermometers which used 0 degrees for the freezing point of water and 80 degrees for the boiling point. Anders Celsius (1701 to 1744) worked professionally as an astronomer, but is best remembered for his centigrade thermometric scale, now known by his name.

¶ Although thermometric expansion was understood well enough for the manufacture of thermometers, it was not until Joseph Black (1728 to 1799) introduced the distinction between temperature and "quantity of heat" that a true understanding of the theoretical nature of heat began to be formulated. He reasoned that heat is a fluid—

Jesse Ramsden (1735 to 1800) constructed a number of important physical instruments, especially of an optical nature. *(From an engraving in the author's library)*

the caloric—which is indestructible. When a body is heated it accumulates caloric. When it cools it loses caloric. The unit of the quantity of heat is the amount of caloric needed to raise the temperature of a given body of mass m by 1 degree C. This quantity is its capacity for heat (now called the specific heat).

¶ Investigations were also undertaken of the conductivity of heat through various substances. By 1760, the concept of latent heat was introduced. Although some hypothesized that the calorific capacity of bodies may be varied by rubbing, by the end of the eighteenth century it was shown conclusively that heat is subject to the general laws of the conservation of energy. The kinetic theory of gases was begun in the eighteenth century, but it would be well into the nineteenth century before this theory could be established conclusively.

¶ The application of these theoretical studies of heat was soon forthcoming. In the mid-eighteenth century, James Watt (1736 to 1819) overcame many of the deficiencies of existing steam engines; this led to better sources of power and the steam engine became the foundation of the modern technological world. Watt also took an interest in chemistry and suggested that water, rather than being a simple substance, is a compound of hydrogen and oxygen.

¶ By the late seventeenth century, electrostatic machines had been invented. However, advances in magnetism and electricity remained sporadic until near the end of the eighteenth century. Among these early advances were studies of electrical conduction and induction, and the discovery of two kinds of electricity. Benjamin Franklin (1706 to 1790) evolved a theory of electricity as a single "fluid." In 1785, Charles A. Coulomb (1736 to 1806) described a method for measuring electrical forces. As the eighteenth century ended, Luigi Galvani (1737 to 1798) and Alessandro Volta (1745 to 1827) described electrical currents and an electrical battery. This represented a clear break with the past, but the consequences were not fully developed until electrolysis and electromagnetism were discovered.

¶ Similar progress was scored in studies of mathematics, chemistry, physiology, geology, and the other sciences. It is of interest to note the contribution of Denis Diderot (1713 to 1784) in this regard. After graduating from the University of Paris, Diderot mixed with the *cafe intelligentsia* of Paris, which included Jean-Jacques Rousseau. Diderot supported himself as a publisher's hack, translating various works into French. This sparked a desire to prepare a work of encyclopedic dimensions. He enlisted the help of Jean le Rond D'Alembert (1717 to 1783), one of the revolutionary thinkers of the period and an extensive contributor to mathematics and mechanics. D'Alembert, however, resigned after the work was partially published, after an attack by Rousseau regarding one of his articles. Nevertheless, the *Encyclopedie ou Dictionnaire Raisonee des Sciences, des Arts et des Metiers*, published in twenty-eight volumes between 1751 and 1772, is regarded as the literary monument of the Enlightenment. It sought to present knowledge as a unified whole and placed emphasis on the progress of the human mind. The work was attacked by the Establishment, particularly in clerical circles, so that the publisher removed some of the more inflammatory passages. However, the work that remained is a landmark in the progress of science, technology, and human development.

Arago's Adventures

Dominique Francois Jean Arago

HE PATHS TO GREATNESS are as numerous and diverse as the roads to Rome. This is typified by the career of Dominique Francois Jean Arago (1786 to 1853). Although one might argue that his name may be excluded from the list of greats in optics, I contend that his researches, his lectures, and above all else, his influence on other investigators, mark him as one who has had a profound leverage on our understanding of optical phenomena. Before we turn to these achievements, however, it is of interest to review his early life.

¶ Arago was born on February 26, 1786, in the commune of Estagel (a department of the eastern Pyrenees). Although he learned the rudiments of reading and writing at the local primary school, his early education seems to have been influenced more by his home and environment. At the time of Arago's childhood, the Spanish and the French were in conflict, and Estagel became a rest stop for French troops. Accordingly, Arago's home was constantly full of soldiers, exciting the youngster toward a military career. His parents counseled him otherwise.

¶ At the age of nine, Arago moved with his family to nearby Perpignan, where he continued his classical education. One day he noted an officer of Engineers who was directing some repair work. Since this officer was quite young, Arago approached him and inquired how he had succeeded in wearing an epaulette at his age. The officer answered, "I come from the Ecole Polytechnique," and explained to Arago that information regarding the entrance requirements could be found at the local library. After obtaining this information, Arago abandoned all consideration of literary studies and devoted himself entirely to a mathematical course. To do this he sent to Paris for works of Euler, Lagrange, and Laplace. He mastered these works on his own.

¶ At the age of sixteen, Arago traveled to Toulouse for the entrance examination in company with a candidate who had prepared at the local college. The companion flunked. Arago, however, answered the questions to perfection, secured the first place, and entered the polytechnic school at the end of 1803.

¶ Indicative of Arago's character is an interview he had with Legendre, who at that time was on the examining board of the school. Legendre opened the interview sharply, "What is your name?" "Arago." "You are not French then." "If I was not French I would not be before you." "I maintain that he is not French whose name is Arago." "I maintain on my side that I am French, and a very good Frenchman too." Legendre relented and sent the young scholar to the board to solve a problem requiring the use of double integrals. Another argument ensued as Arago used a technique developed by Legendre himself. Legendre asked if this method had been chosen as a bribe. Arago replied, "Nothing was further from my thoughts. I only adopted it because it appeared to me to be preferable." Arago continued in this indomitable way and finally found Legendre to be satisfied and to relent. Later they became good friends and colleagues.

Dominique Francois Jean Arago as permanent secretary of the Academie des Sciences influenced much of the European optical advances being made in the post-Napoleonic era.
(From an engraving by John Sartain, copy in the author's library. Sartain probably engraved this for the Eclectic Magazine *of 1854.)*

¶ During the last months of Arago's attendance at the school, there was considerable unrest due to the requirement for students to publically support Napoleon. Many students were dismissed because of insubordination; Napoleon could not, however, censure Arago who sat at the head of the class.

¶ Upon graduating, Arago accepted a position at the Observatory. Soon thereafter he met Jean Baptiste Biot (1774 to 1862) with whom he discussed the interest to extend the meridional line further south. This determination would permit a more precise determination of the size of the earth and, hence, the meter. The plan was submitted to Laplace who received it enthusiastically and procured the necessary funds. In 1806 Arago, Biot, and a Spanish commissary departed from Paris.

¶ Arago's first stop was in Spain, where he met the daughter of a Frenchman. As all hotels were crowded, she invited Arago to take refreshments at her grandmother's home. Upon leaving, Arago was informed that the French lady's betrothed was in disfavor of this visit and that Arago should be prepared for an attack. Accordingly, he purchased some pistols. On his return trip, Arago confided to his driver that they might be stopped but not to worry for he had a pistol to protect them. To which the driver retorted, "Your pistols are completely useless; leave me to act." Two men did attempt to halt Arago's carriage, but a cry from the driver was sufficient to cause the mule to rise up almost vertically, raising one of the assailants from the ground, then throwing him to the ground. They then departed at a rapid gallop. The next day Arago was informed that a man had been crushed to death on the road he had passed.

¶ Arago set out to establish a station on top of a mountain in the area so that he could triangulate over a considerable distance. Despite being informed that this area was full of highway robbers, Arago set out. The first night at this station the rain fell in a deluge. Towards night a stranger sought shelter in the cabin, and Arago permitted him to share the comfort of his cabin. This stranger was the chief of the local bandits. A few days later he sought shelter again, upon which Arago's servant determined to kill the man. Arago intervened, stating, "Are you mad? Are we to discharge the duties of the police of this country?" This was a fortunate decision. The robber band enabled Arago to travel from one of his stations to another safely.

¶ Arago went on to Majorca where he established another mountain-top station. A rumor spread that his purpose was to spy and send signals to the French army. Arago abandoned the station, seeking refuge across the harbor via a small boat. He was hotly pursued but, fortunately, arrived on shore with but minor wounds and headed for sanctuary in the dungeon of the castle of Belver. Arago noted that prisoners had often fled *from* this dungeon and that he was perhaps the first to have this happen in reverse.

¶ While in this self-imposed incarceration, Arago read a journal describing the execution of M. Arago. He reasoned that this conjecture might soon turn to fact, so he fled. A few weeks later, as he neared Marseilles, he encountered a Spanish corsair. He was taken prisoner again, and his life seemed to hang by but a thread. Writing to the captain of a nearby English vessel, he requested that he be taken aboard. He even suggested that if the captain could not grant this favor, to at least take Arago's papers to the Royal Society. The captain refused. Some time later, this information was received in England and the captain was severely reprimanded.

¶ After nearly four months as a prisoner, Arago again escaped and set sail for Marseilles. He was unable to land there, but after another month at sea, he arrived at Algiers. Seven months later he finally returned to France. After completing his quarantine he was reunited with his family in Perpignan.

¶ Returning to academic life after three years during which one is feared dead is not an easy handicap to overcome. Let us see how Arago fared.

¶ It is important at this juncture to understand the position of science in France at the beginning of the nineteenth century. The Academie des Sciences had been dissolved in 1793 and replaced two years later by the Institut de France. This was divided into three sections or "classes"; the first class was that of mathematical and physical sciences, or the Academie. Membership in the first class could be obtained only when an appropriate vacancy occurred.

¶ Upon the death of Lalande on April 4, 1807, a vacancy was available in the first class. Six weeks later (a decent period after the death of a distinguished astronomer) the first class considered the request to fill the vacancy, but by unanimous vote it was decided to postpone the election for six months in hopes that Arago would return. On December 7, 1807, it was decided to postpone the election again, although by now it was feared that Arago was lost and likely dead.

¶ On September 30, 1808, Arago's father informed the Academie that Francois was alive, and within a year he arrived in Paris. On August 28, 1809, Arago was welcomed back and allowed by special invitation to take part in a meeting. Soon thereafter he read a memoir on his work in Spain. Within a fortnight he was elected to Lalande's vacant chair.

¶ It was a great honor for Arago, at the age of twenty-three, to be so saluted. He immediately took an active interest in the politics of the Academie and was elected permanent secretary in 1830. Arago was also made director of the Paris Observatory and professor of descriptive geometry at the Ecole Polytechnique in 1809. He was co-editor with Joseph L. Gay-Lussac (1778 to 1850) of the *Annales de Chemie et de physique* from 1816 until 1840. On the initiative of Arago, the Academie began publication of the *Compte Rendus Hebdomadaires des Seances de l'Academie des Sciences* in 1835. He received the Royal Society's Copley Medal in 1825.

¶ Arago's most important original work in optics was carried out prior to 1830. At that time he became involved in politics, being repeatedly elected to the Chamber of Deputies where he sat on the left. He delivered influential speeches on educational reforms, freedom of the press, and the application of scientific knowledge to technological progress. His political career peaked in 1848 when he was made a member of the provisional government and was named minister of the navy and the army and president of the Executive Committee. As minister he signed decrees outlawing corporal punishment, improving rations for sailors at sea, and abolishing slavery in the French colonies. He was passionately concerned with social reform and the education of the lower classes.

¶ Arago's contributions to optics involve the establishment of the wave theory of light, investigations of polarization phenomena, studies of atmospheric scintillation using interference phenomena, and studies of atmospheric refraction. However, Arago is perhaps best remembered for the encouragement and inspiration he gave to others. Francois Arago is certainly one of the greats of optics.

The Napoleonic Influence

THE PATRONAGE OF SCIENCE by ruling governments began in the days of antiquity. However, the influence of the ruling power on the development of science is particularly marked by the career of Emperor Napoleon Bonaparte. Napoleon was born in Corsica on August 15, 1769, and attended military school there. He was regarded as taciturn and morose, but since he spoke little French in this French-speaking environment, this might have been expected. He distinguished himself in mathematics, did tolerably well in history and geography, but did poorly in Latin and general literature. In 1784 he left for Paris, where his excellence in mathematics permitted him to finish his military training and receive a commission a year later instead of in the two years normally required.

¶ This was a period of accelerated social upheaval. The revolution in America may be regarded as the direct cause of the French Revolution in 1789. Political dissensions were combined with a severe economic depression. As if that weren't enough, the winter of 1788/1789 was the harshest of the century.

¶ Early in 1792, Napoleon was promoted to captain of the artillery. Within a year he was promoted to lieutenant colonel and he then began to display his military genius. He was then promoted to brigadier general. In October 1795, Napoleon, with the command of 5,000 troops, secured Paris against a foe of 30,000. For this he was appointed to the command of the Army of the Interior. The revolution had effectively ended with the power of the French government in the hands of a military adventurer.

¶ It must be understood that although Napoleon was primarily a militarist, he had a good grasp of mathematics and a strong interest in science. Early in his career, he befriended Gaspard Monge (1746 to 1818), the founder of descriptive geometry, and Claude Berthollet (1748 to 1822), the eminent chemical innovator, among other savants. As a result, Napoleon was elected to membership in the Institute (as the Academie was then called), prior to his coming in power. Napoleon savored this distinction and throughout his career he supported science, in contrast to his apparent contempt for the arts.

¶ Napoleon coupled his military achievements with scientific activity. In 1798 he set out to conquer Egypt, taking with him an army of 36,000 and a body of scientific and learned men. His military objective was to maintain a route to India, but he also avowed to bring European erudition to the heathen. To do this he attempted to establish an Institute of Egypt. This effort was not entirely successful.

¶ He participated actively in academic affairs early in his career. In the winter of 1801/1802 the great Italian physicist Alessandro Volta (1745 to 1827) visited Paris and delivered three lectures to members of the Institute. Napoleon attended all three. Moreover, he then established a prize to encourage anyone to produce a scientific advance comparable to those of Franklin and Volta. The announcement of this prize was made with great fanfare, but despite the obvious encouragement, the prize was not awarded until five years later. This was also the last time that Napoleon attended meetings of the Institute (but it might be noted that other interests took an increasing amount of his time).

¶ Napoleon also supported many other scientific efforts. One that he did not support is of interest. Robert Fulton (1765 to 1815), the American pioneer in steam propulsion of boats, undertook tests of his submarine *Nautilus* on the Seine in 1800. Confident that this invention would be a crucial weapon against the British navy, Fulton sought protection to cover the possibility that he might be caught and tried as a spy. Napoleon appointed Laplace and Monge to examine the submarine. They did and recommended that Fulton be given a government grant. This was awarded and then withdrawn when it appeared that the craft would not be of immediate use since means for self-propulsion had not yet been developed. Three years later, Fulton returned with his concept of a steamboat. Napoleon felt that such an invention "may change the face of the world" but the French scientists were unimpressed. Reflect that Fulton received less encouragement from King George III and from President Thomas Jefferson.

¶ In 1804 Bonaparte was given the title of Emperor. He requested that Pope Pius VII perform the ceremony of coronation, but the Pope was allowed to perform only part of the ritual. Napoleon snatched the crown from the Pontiff's hand and placed it on his own head.

¶ Napoleon was passionately admired as well as being disliked. Many liberals among the scientists, for instance, resisted his power. On the other hand, men such as Nicholas Chauvin, a soldier of the French Republic and of the First Empire, had such an idolatrous respect for the Emperor that today a feeling of exaggerated devotion or patriotism is often called "chauvinism."

¶ Napoleon encouraged foreign scientists to reside in France and frequently offered them stipends for this. He encouraged industry with subsidies and loans as well as the cheaper reward of honor. As Emperor, Napoleon had less time to devote to science but he still took counsel from the scientists. He read scientific books, but had insufficient time to digest them. Of course, in 1814 he had abundant leisure on the island of Elba. There he took with him the textbook on physics written by the abbe Renatus Hauy (1742 to 1822). Bonaparte had urged the abbe to write this text and to regard education as a matter of great importance. When Napoleon undertook his final period of enforced leisure on Saint Helena in 1815, he took with him about a dozen references on science, including George Buffon's 127-volume *Histoire Naturelle*, Delambre's astronomy text, and Fourcroy's treatise on chemistry.

¶ Bonaparte's support of science was not always consistent, but his patronage of several scientists permitted French science to blossom profusely in the nineteenth century. Of particular interest is his patronage of his friend, Claude Berthollet. Berthollet had been sent to Italy in 1796, where he met Napoleon. A bond of friendship soon arose and the chemist was asked to participate in many of Bonaparte's activities, including the scientific commission that accompanied the expedition to Egypt. Bonaparte came to regard Berthollet as "my chemist" and bestowed honor and patronage upon him. Soon after returning from Egypt in 1799, Berthollet bought a country home in the village of Arcueil, about three miles south of Paris.

¶ In 1806 Pierre Laplace (1749 to 1827), a distinguished scientist and a friend of

Napoleon, purchased the property adjoining that of Berthollet. With their well-paid positions in government it was possible for them to establish a private scientific society. The Society of Arcueil came to be the most important scientific organization of the age. It complemented the Institute by providing laboratory facilities not available there. The Society restricted itself to the study of physics and chemistry, and their contributions to optics and physical chemistry are particularly noteworthy.

¶ The original members of the Society included Jean Baptiste Biot (1774 to 1862) and Joseph Gay-Lussac (1778 to 1850). In 1804 they made a daring balloon ascent together to an altitude of 23,000 feet to obtain samples of the air. Analysis proved that it had the same composition as that near the ground. Biot later joined Arago on an expedition to Spain to measure the arc of the meridian and thus to establish the length of the meter. Biot's most influential work was in the extension of Arago's investigations of the phenomenon of rotation of the plane of polarization in certain materials. Polarimetry as an analytical tool sprang from these researches.

¶ Etienne Malus (1775 to 1812) joined the Society soon after it was formed. He participated in Napoleon's Egyptian expedition where he discovered a new branch of the Nile. He later investigated double refraction in crystals which led to investigations important in establishing the wave theory of light.

¶ Francois Arago (1786 to 1853) joined the Society at an age of but twenty-one. His contributions to optics are legend. Others who were elected to the Society include the chemical encyclopedist Jean Chaptal (1756 to 1832), the physical chemist Pierre Dulong (1785 to 1838), and the physicist Simeon Poisson (1781 to 1840). The contributions and influence of these men established France as the leader of scientific investigations in the nineteenth century.

¶ It would be an oversimplification to credit Napoleon with the emergence of this cadre of capable scientists. On the other hand, it would be chauvinistic to ignore his contributions. ☙

An Unexpected Discovery

Etienne Louis Malus

OUBLE REFRACTION IN A CRYSTAL had been observed as early as 1669 by the Danish naturalist Erasmus Bartholin (1625 to 1698). Using a crystal of Iceland spar he showed that an incident beam of light was split into two rays by what he called ordinary and extraordinary refraction. He could give no theoretical explanation for this. Christiaan Huygens (1629 to 1695) advocated a wave theory of light that would explain double refraction as the formation of spherical and spheroidal waves. However, since Huygens considered the waves to be longitudinal rather than transverse oscillations, his wave theory could not account for many of the other observations of light phenomena, hence was not generally accepted. The readily observed phenomena of double refraction thus remained an enigma into the eighteenth century.

¶ Today, of course, interpreting double refraction presents no problem. Using an analyzer, one readily discerns that the two rays are polarized in mutually perpendicular planes. To a seventeenth-century observer, the concept of an analyzer was not commonplace. The discovery of the fact that the rays are polarized is a remarkable story which involves a remarkable man.

¶ Etienne Louis Malus was born on July 23, 1775, in Paris. His early training was principally literary; he acquired a sound knowledge of Greek and Latin authors. When it came time to enroll in a school of higher learning, the French Revolution was rampant. Consequently, Malus disdained further education and enlisted in the 15th Battalion of Paris and was sent to Dunkirk to assist in the construction of field fortifications.

¶ It was at Dunkirk that Malus first demonstrated his latent talents for technical creativity. The engineer who was directing the constructions found that the soldiers were executing the activity in a manner other than as they had been directed. Upon inquiry he found that Malus was responsible for this alteration in procedure. Malus had proposed a technique which accomplished the purpose with the least possible fatigue. Realizing that Malus had abilities superior to the tasks being assigned, the engineer sent him to the Ecole Polytechnique, which had just been founded.

¶ Upon completing his schooling, Malus returned to the army, this time as an officer. Two years later, he was sent to Egypt to participate in Napoleon's actions there. Malus scrupulously narrated, in a journal, every event to which he was a witness. He described the battles which took place. Of greater interest here are his descriptions of reconnoitering the Nile to determine more precise information regarding distances to the sea. Here he described his discovery of the ancient city of San, or Thamis. He later made other discoveries of interest to archeologists about the ancient geography of the upper Nile valley.

¶ Malus then took part in the expedition to Syria, during which the soldiers traveled with insufficient provisions and drinking water. At Jaffa their seige met with keen resistance. A sortie of troops surprised the French, carrying back the heads of the victims, for which they were paid by their weight in gold. Fortunately for Malus and for optics he, by reason of being asleep in one of the trenches, escaped this fate.

¶ Later, the French did occupy Jaffa. Malus's descriptions of this reflect the outrage of warfare: *The tumult of carnage, the broken doors, the houses shaken . . ., the cries of the women, the father and child overthrown one on the other, the violated daughter on the corpse of her mother . . . was the spectacle which this unfortunate city presented.* The army then set out for further advances, leaving Malus with but 150 men. Meanwhile, the plague infected those left. For ten days Malus passed in this environment with no sign of direct infection. On the eleventh day he wrote, *A burning fever, and violent pains in the head, forced me to seek repose . . . and one by one the symptoms of the plague showed themselves. . . . I resigned myself to my fate.*

¶ At length Malus was placed on board a vessel setting sail for Egypt. The captain had the plague and died enroute. But Malus found the sea air beneficial; his strength returned; the inflammation in his groin left; and his appetite was restored. Reaching Egypt, he was placed in quarantine along with numerous people suffering from the plague. Many died, and Malus provides a grisly account of this. After about two months of this misery, Malus was released.

¶ This was hardly the environment one would prescribe in which to write a memoir on light. But Malus did that, and we shall discuss its relevance shortly. After engaging in several more battles, Malus finally returned to France, arriving in October 1801. Although he was again quarantined, he must have found that he was now confined in comparative luxury.

¶ Returning to Paris, Malus married the girl to whom he had been affianced for four years. His subsequent army career is of but minor interest.

¶ The memoir on light which had been written in Egypt proposed that light is constituted as a mixture of oxygen and caloric in a particular state of combustion. The different natures of light were attributed to the proportion of caloric they contained. Malus essentially subscribed to the corpuscular theory. While this memoir may not be important to our understanding of the nature of light, Arago pointed out that *no army in the world ever before counted in its ranks an officer who occupied himself in the spare hours of advanced posts with researches so complete and so profound.*

¶ Malus presented this and two more reports to the Institute in 1807; the second treated rays of light passing through an optical system in three dimensions; the third described a method of determining the refractive power of opaque objects. None of these is monumental, but a paper delivered in 1808 placed Malus's name among those who have made important contributions to optics.

¶ Malus, at the time, was residing in the Rue d'Enfer within sight of the Palace at Luxembourg. One afternoon he happened to examine, through a doubly refracting crystal, the rays of the sun reflected from the palace. Instead of the two bright images which he had expected to see, he perceived but one. By rotating the prism, first the ordinary ray and then the extraordinary ray was transmitted. Malus was astounded at this and at first attempted to explain it by supposing that some modification imposed by the atmosphere in transmitting the solar light was responsible.

¶ That evening he caused the light of a candle to be reflected from the surface of water at an angle of 36 degrees (being the angle the ray makes with the horizontal) and found that this light, too, was polarized. He repeated the experiment with a glass plate and found that at 35 degrees this light was polarized. Malus thus discovered that polarization was caused by means other than by double refraction. Malus later extended his observations to show that light which first passes through a doubly refracting crystal and is then reflected will transmit but one ray.

¶ Malus extended his observations to show that natural light can be partially polarized. He also showed that light refracted and then reflected from a pile of glass plates will result in rays which are polarized mutually perpendicular.

¶ His contribution was quickly acknowledged. He was named a member of the Society of Arcueil, and in 1810 he was elected to membership in the Institute. Unfortunately, the seeds of his mission in Egypt and Syria soon bothered him. In mid-1811 the symptoms of consumption made alarming and rapid progress. He died that year with his head reposing against that of his wife.

¶ I would like to end this anecdote regarding the discovery of an ability to analyze polarized light with one of the maxims that Malus considered appropriate:

¶ *I will found my enjoyments on the affections of the heart, the visions of the imagination, and the spectacle of nature.* ☙

The Legacy of Fourier

IN HIS EULOGY OF JOSEPH FOURIER to the Paris Academy of Sciences, Francois Arago concluded, "My object will have been completely attained if . . . each of you have learned that the progress of general physics, of terrestrial physics, and of geology will daily multiply the fertile applications of the *Theorie de la Chaleur*, and that this work will transmit the name of Fourier down to the remotest posterity." Although Arago thus predicted a legacy of the use of Fourier's mathematical treatment used to describe the conduction of heat, in 1833 he could hardly envisage those benefits to be derived in optics. Today, Fourier optics and Fourier transform spectroscopy are widely practiced by scientists with little knowledge of the life of Joseph Fourier.

¶ It is of interest to review Fourier's life and to reflect on the manner of man who brought to us a keener understanding of optical phenomena. Much of my information about his life is derived from the aforementioned eulogy, which was written by a friend and admirer. It was also written after a turbulent period in French history when the passion of patriotism gave rise to overindulgent attitudes toward behavior. I cite this to indicate that I have tried to separate fact from fancy, and to warn you that my attempts to present a clear account of Fourier's life may, at times, be colored by the biases introduced into the documents available to me.

¶ Fourier was born at Auxerre, about 160 kilometers southeast of Paris, on March 21, 1768; he was orphaned at the age of eight. A neighbor lady, who recognized his courteous manners and his precocious natural abilities, recommended him to the Bishop of Auxerre, and Fourier was admitted to the military school conducted locally by the Benedictines. His precocity was soon evident, as Fourier anonymously authored many of the sermons delivered by high dignitaries of churches in Paris.

¶ Although evidently a gifted child, Fourier was also petulant, noisy, and vivacious. However, during his fourteenth year he became interested in mathematics and settled down (or as Arago writes, "He became sensible of his real vocation.").

¶ Educated in a military school directed by monks, Fourier wavered between a career in the church and with the military. He preferred the latter, but this was not then possible because Fourier's father had been a tailor and not of the nobility. Fourier thus entered an abbey, but before taking his vows the social upheaval in France attracted him to a teaching position. He was appointed to the principal chair of mathematics in the Military School of Auxerre.

¶ He soon displayed an unusual talent for lecturing on rhetoric, history, and philosophy, substituting for his colleagues when they became ill. His lectures on these several topics attracted a delighted audience of diverse backgrounds. This characteristic distinguished Fourier throughout his career.

¶ In 1789 Fourier read a paper to the Paris Academy of Sciences on the resolution of numerical equations of all degrees. This work, enunciated by Fourier at the age of twenty-one, formed the cornerstone upon which he developed his future mathematical work. In this paper, Fourier extended some of the prior contributions of Lagrange. Nevertheless, Fourier's accomplishment had little appeal to the pure mathematicians, who felt that it lacked rigor. To the physical scientist, however, Fourier's results were received warmly (and still are) since they simplified calculations.

¶ Returning to Auxerre, Fourier enthusiastically embraced the principles of the Revolution. This was a period of revolution not only politically but in the arts and sciences as well. For example, the reformation of weights and measures was begun at this time, leading to the introduction of the metric system. Alas, the times had taken many of the savants into military activity or, as with Lavoisier, had removed them permanently.

¶ Fortunately, Napoleon in his rise to power realized the futility of ignorance in building a meaningful empire and he encouraged the creation of schools. In 1794, the Ecole Normale was started and Fourier was rewarded for his patriotism in Auxerre by being appointed to the chair of mathematics. This school lasted but a few months, at which time the Ecole Polytechnique was established. Again, Fourier was called and he responded by embellishing his reputation for clearness, method, and erudition. His lectures attracted a fastidious and wide audience.

¶ Soon after the formation of the Ecole Polytechnique, Napoleon began to dream of restoring Egypt to its ancient splendor. And what better way to accomplish this than by introducing French culture to this now-backward country? Napoleon realized that to achieve this goal he would need more than a mere army. He would require leading scientists, and he chose Gaspard Monge and Claude Louis Berthollet. Both of these men were members of the Paris Academy of Sciences, on the faculty of the Ecole Polytechnique, and recognized as being among the leading scientists of the time. They in turn asked Fourier to join them and he did so. In Cairo they established the Institute of Egypt with Monge as the first president and Fourier as perpetual secretary.

¶ In Egypt, Fourier distinguished himself by extending his mathematical researches to a general solution of algebraic equations, methods of elimination, and indeterminate analysis. His breadth of interest was also manifest by his contributions in general mechanics, an aquaduct to conduct water from the Nile to the Castle of Cairo, a proposal to explore the site of the ancient Memphis, a descriptive account of the revolutions and manners of Egypt, and a wind machine to promote irrigation.

¶ However, Napoleon's dream to *rescue them from the galling yoke under which they had groaned for ages . . .* [and] *to bestow upon them without delay all the benefits of European civilization* (I quote from Arago's eulogy) was a failure. The uncivilized Egyptians failed to respond to the cultural feast proferred. Napoleon surreptitiously returned to France, as did Monge, leaving Fourier to cope. During Fourier's continued stay in Egypt, Napoleon conquered much of Europe and became the virtual ruler of France. But his fortunes in Egypt never proved successful and within three years of Napoleon's stealthy departure, Fourier and the others went back to France.

¶ Upon his return to France, Fourier was named Prefect of the Department of l'Isere. Although this area was a hotbed of political dissension, Fourier, with great diplomatic skill, soon established harmony among the near-warring factions. The situation was brought to such a quiet state that Fourier could continue his efforts in mathematics and letters. From Grenoble, the principal city of Isere, Fourier wrote his *Theorie*

Mathematique de la Chaleur. This was Fourier's outstanding scientific effort.

¶ Fourier's effort received a mixed reception. The pure mathematicians again pointed out the lack of rigor in his treatment. Pure mathematicians and mathematical physicists have nearly always been at odds, the former disdaining any treatment that avoids the scrutiny of rigid proof and the latter pleased to have a procedure to express the results of their observations.

¶ Fourier recognized that any function whose graph displays a periodicity can be considered to be a sum of sinusoidal functions. That is

$$f(x) = \sum_n A_n \sin nx.$$

(The purist may note that I have taken some liberties in expressing the Fourier series. Pshaw.)

¶ This series is now known as a *Fourier series*. Its real value to optics, of course, is that it leads to an integral transform whereby a periodic function of space, for example, may be transformed to a periodic function of time. By this means the spectral characteristics of a radiating source may be separated by a Michelson interferometer to provide a time-varying signal in which time is not directly related to wavelength. However, by means of the Fourier transformation this can be accomplished. Generally, the transformation is undertaken with a computer, although before computers other means were employed.

¶ Fourier had submitted his treatment of the conduction of heat, in which his series was fundamental, to the Paris Academy in 1811, for which he was awarded its mathematical prize in 1812. As noted, some reservations were expressed with the favorable judgment. However, Fourier never admitted the validity of this dissension, giving unmistakable evidence near the close of his life that he thought it still unjust by causing this memoir to be published in the Academy records without changing a single word!

¶ This work gave a tremendous impetus to the research of his colleagues who considered the geological heat content, the temperature of celestial regions, and the effects of heat on biological growth. During this period, Napoleon's influence had blossomed and faded. In 1815, Napoleon escaped from Elba and made a triumphal march on Paris. Fourier had mixed reactions to this news. He left Grenoble for Lyons, where some of the royalty had assembled. They greeted Fourier coldly and doubted that Napoleon could have captured nearby Grenoble. Consequently, Fourier was told to return and protect the (already fallen) city. Fourier had barely left Lyons when he was arrested by Hussars and conducted to Napoleon's headquarters. Fourier explained that his *duty* compelled him to act as he had. Napoleon forgave Fourier, but did not endear himself when he told Fourier, "I have made you what you are."

¶ Fourier was appointed Prefect of the Rhone and given the title of Count—promotions Fourier dared not reject. However, this appointment as Prefect lasted but a short time. Fourier returned to Paris with no income and no financial reserve. It was a turbulent time for many. Napoleon's career ended at Waterloo, and the Bourbons were restored to power in Paris.

¶ Fourier applied for a federal pension for his fifteen years of service to his country. He was rudely repelled. However, a former student at the Ecole Polytechnique, upon learning of Fourier's plight, enabled him to receive the directorship of the Bureau de la Statistique of the Seine.

¶ The Academy of Sciences sought at its first opportunity to elect Fourier to the society. Political intrigue, sanctioned by Louis XVIII, prevented anyone who had been associated with Napoleon from election to the Academy. (Arago noted, "In our country, the reign of absurdity does not last long.") A year later, in 1817, the Academy again unanimously nominated Fourier to a place in the section of physics. This time there was royal confirmation without difficulty.

¶ Fourier was now able to spend the last years of his life in retirement and in the discharge of academic duties. He became eloquent in discoursing on those facets of life which he had experienced. There are those who find this type of eloquence somewhat boorish, rather than fascinating. They cite an incident in Fourier's later years as a case in point. Fourier was seated at a table together with some who were strangers to him. One,

Joseph Fourier, French mathematician. From a sketch by Boilly. *(The Bettmann Archive, Inc.)*

in particular, was identified as an old officer. To him, Fourier described in great detail the events of a battle that had taken place in Egypt, of which Fourier had some first-hand knowledge. Fourier concluded his recitation of the details of this battle by noting, complacently, that he felt his memory had served him correctly in recalling these events. His companion, who seemed to have been enthralled by this discourse, assured Fourier that his statements were accurate, and that he based this judgment on the fact that he, too, had personal knowledge of the battle, having been head of the Grenadiers involved!

¶ Although endowed with a sturdy consititution, Fourier had adopted the habit of wearing too much clothing. Thus, although he gave the appearance of corpulence he was, in fact, a quite slender man. He abided in a sterile, ovenlike environment, even keeping his windows closed in the heat of summer. Visitors found this to be annoying. As a result of this, Fourier developed an aneurism of the heart. In the spring of 1830 he further sustained a fall while descending some stairs. This aggravated his condition and within two weeks he died.

¶ Fourier's name is now used as an adjective describing an eloquent method of handling several optical processes. Thus, we who work in the field of optics revere this man, perhaps unaware that the legacy he left us was beyond expectation. ☺

Young and the Rosetta Stone

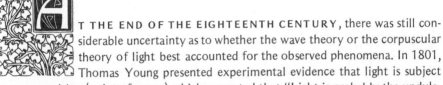 T THE END OF THE EIGHTEENTH CENTURY, there was still considerable uncertainty as to whether the wave theory or the corpuscular theory of light best accounted for the observed phenomena. In 1801, Thomas Young presented experimental evidence that light is subject to superposition (or interference) which suggested that "Light is probably the undulation of an elastic medium." This hardly put the matter to rest, as we shall see. Moreover, the resulting controversy brought so much turmoil into Young's life that he tended to avoid further research regarding the nature of light. However, his genius was not extinguished. Let us examine some of the details of his life.

¶ Thomas Young was born at Milverton on June 13, 1773. At that time Milverton was a quiet village in Somerset, England (and still was during my visit there in the mid-1960s). Thomas was a precocious child who, by the age of two, read with considerable fluency. Before he was four he had read the Judeo-Christian Bible twice. By the age of six, he had begun the study of Latin grammar. Soon thereafter, he began a study of Italian and French. Later he studied various middle-Eastern languages, including Syriac and Hebrew.

¶ He also displayed considerable mechanical skills. At the age of thirteen, Young had learned the use of the lathe, the making of telescopes, the grinding and preparation of colors, and the binding of books. He used these skills as a source of income with which he purchased Greek and Latin books.

¶ His competence in language is evident in a story regarding a visit he made to London with an aunt. There, in one of the bookseller's stalls, they found a valuable classic which attracted Young's attention. The owner of the shop, thinking that the quaintly dressed Quaker lad was indulging in ignorant curiosity, offered to make a gift of the book to the boy if he could translate one page. To the man's astonishment, Thomas rapidly turned the text into flowing English. The bookseller winced at his (sacrificial) offer, but made the boy the proud possessor of a valuable book.

¶ By the age of seventeen, Young had mastered Newton's *Principia* and *Optics*. However, he was persuaded to study medicine and at nineteen he began a medical education in London. This education was continued at Edinburgh and Gottingen, and finally at Cambridge. In the spring of 1799, Young entered medical practice in London.

¶ During his medical education, Young undertook research on vision and submitted a paper to the Royal Society regarding the functions of the crystalline lens of the eye. This was an important contribution to an understanding of accommodation. Young developed an optometer which was used to measure the length of the optic axis of the eye and the radius of curvature of the cornea. Using this instrument, Young made the first recognition of astigmatism, as follows:

> *My eye, in a state of relaxation, collects to a focus on the retina, those rays which diverge vertically from an object at the distance of ten inches from the cornea, and the rays which diverge horizontally from an object seven inches distance. For, if I*

The town of Milverton commemorates its most illustrious son, Thomas Young, with this etching hung on a coat rack in the local school. Here, Mr. Hoare, the Headmaster, examines some books alongside this memorial. *(Photograph by the author)*

hold the plane of the optometer vertically, the images appear to cross at ten inches; if horizontally at seven. I have never experienced any inconvenience from this imperfection, nor did I discover it till I made these experiments; and I believe I can examine minute objects with as much accuracy as most of those whose eyes are differently formed.

¶ Young next studied color perception, noting first that the eye suffers from chromatic aberration. From this he went on to suggest that perhaps the retina contains three types of perceptors, each capable of vibrating at some fixed frequency, but being able to be put into motion by the undulations of light striking the retina. Thus, Young conjectured that if the receptors are sensitive to three principal colors—red, yellow, and blue—that green light will excite equally the yellow and blue receptors. He later revised his principal colors as red, green, and blue, asserting that a yellow sensation can be caused by a mixture of green and red light.

¶ At about this time Benjamin Thompson (1753 to 1814) was establishing the Royal Institution in London as a public institution where scientific knowledge could be extended and applied to "The Common Purposes of Life." Thompson was a recalcitrant who fled America at the time of the Revolution because of his Royalist sympathies. He left England after disagreements arose there. In Bavaria he was made Count Rumford as a result of his investigations of heat, which were inspired by a need to economically prepare meals for the poor of Munich. With his reputability reasonably restored, he sought to redress some of his former capricious actions by establishing the Royal Institution. (He later was involved in establishing the Military Academy at West Point and nearly became its first commandant.)

¶ In 1801 Young was hired as a professor of Natural Philosophy at the Royal Institution. His ability as a lecturer did not attract appreciable audiences and Young soon resigned that position.

¶ By this time Young had reported his "double-slit" experiment and his advocacy of the wave theory to explain the observed interference pattern. Before discussing some of the reaction to this pronouncement, it is appropriate to review an exchange of notes which were published and concern Young. In 1800 Young published a paper dealing with the analogy of light and sound, and suggesting that light was, perforce, a wave motion. Anticipating objections to explain interference on this basis, he made a somewhat cavalier remark about the ability of others to correctly handle acoustic harmonics. He was taken to task for this by the botanist and mathematician John Gough (1757 to 1825) who, although he praised Young in expectation of greater achievements, mildly upbraided him for some derogatory remarks he had previously made. Young, for reasons that escape me, had vilified "a young gentleman from Edinburgh" who "fancies he has made an improvement of consequence, when, in fact, he is only viewing an old object in a new disguise."

¶ These comments were unfortunate because, although possibly justified, they carried a tone that caused lasting resentment. The writer he had vilified was young Henry Brougham (1778 to 1868). In 1802 Brougham took revenge by writing of Young's Bakerian Lectures (which described the double-slit experiment):

As this paper contains nothing new which deserves the name, either of experiment or of discovery, and as it is, in fact, destitute of every species of merit, we should have allowed it to pass among the multitude of those articles which must always find admittance into the collections of a society which is pledged to publish two or three volumes every year.

¶ (The publication he was referring to was the highly respected *Philosophical Transactions* of the Royal Society.) Brougham continued his protest by condemning the Royal Society for even recognizing Young's "lucubrations."

¶ Twice more Brougham castigated Young's work. Young was stung, for these articles damaged his reputation both as a physicist and as a physician. Although the audience to which Brougham wrote had no way to know that he had no qualification to make these criticisms, and further they could not know that Young's experimental procedures were beyond redoubt, the articles carried conviction and the damage was done. Young attempted a reply, but it merely muddied the waters. Young retreated from physics.

¶ About the same time that Young made his investigations of light, French troops in the course of some excavations near a little town called Rosetta in the Nile delta uncovered a stone with writing in three different scripts. The potential of this discovery was obvious: if, as seemed probable, the subject matter of all three inscriptions was the same then it might provide the key to interpreting the Egyptian hieroglyphics.

¶ The stone was surrendered to the British in 1801 and taken to the British Museum, where it still resides. Impressions of the inscriptions were made and distributed to scholars all over Europe. Unfortunately, the slab of black basalt was found in a severely damaged condition, leaving part of the hieroglyphics missing. This, of course, hindered attempts in interpreting the hieroglyphics. Several scholars attacked the problem with but marginal success.

¶ It was not until 1814 that Young became interested in the Rosetta stone. Within a few years he had made sufficient progress to publish his method of interpreting the hieroglyphics. Meanwhile, Jean Francois Champollion (1790 to 1832) had been struggling with the translation. In 1821, two years after the last publication of Young's effort, Champollion published an account repudiating Young's interpretations of the hieroglyphics. A year later, Champollion published an account in which his researches now agreed with Young, but he did not acknowledge Young's efforts. Today, one generally regards Champollion as being responsible for the translations. After all, he was a professional Egyptologist and he did carry the translation beyond the point reached by Young.

¶ On a visit to Milverton nearly twenty years ago I was struck by the fact that few people I met there realized that I, an optical engineer, might be interested in their illustrious townsman. After all, his fame arose from unraveling the mystery of the Rosetta stone, which is far afield from optics! However, the inscription on the tablet commemorating Young at Westminster Abbey reads in part:

A man alike eminent in almost every department of human learning . . . [who] *first established the undulatory theory of light, and first penetrated the obscurity which had veiled for ages the hieroglyphics of Egypt.*

Confirmation of the Wave Theory

Augustine Jean Fresnel

 HE NINETEENTH CENTURY had hardly begun when Thomas Young (1773 to 1829) fanned the fire smoldering between supporters of the wave and the corpuscular theories of light. Young's demonstrations that light can be superimposed (i.e., cause interference) gave strong evidence for the validity of the wave theory. However, within a decade Etienne Malus (1775 to 1811) discovered that light can be polarized by reflection. It was soon shown that light so polarized can also suffer interference. If light consists of longitudinal waves as Young had thought—and Christiaan Huygens (1629 to 1695) had proposed —how would it be possible for such waves to be both polarized and to interfere? Augustine Jean Fresnel recognized that light must be a transverse wave. His colleagues were slow to understand this, but Fresnel persisted with complete success.

¶ Augustine was born on May 10, 1788, at Broglie, which is located in the province of Normandy. His father was an architect employed in the construction of a fort in the harbor of Cherbourg. The disturbances of the French Revolution caused the elder Fresnel to move to the small nearby village of Caen with his family.

¶ The youngster showed little aptitude toward learning; in fact, by the age of eight he could scarcely read or write. One would hardly have predicted that he would make a mark on the scholarly world. However, his young colleagues dubbed him "the genius" because he was able to determine the length and bore that gave the greatest power to the little elderwood popguns with which the children played. He could also determine the woods which are best to use in making bows in consideration of their elasticity and strength. As a matter of fact, the parents of his playmates prohibited further developments of this kind because the toys had become dangerous arms.

¶ After three years in school at Caen, where he had moved to live with his older brother, Fresnel entered the Polytechnic School. He was now sixteen and a half, but so frail in health that he was not expected to complete his education. However, he was buoyed by praise from men such as Legendre for his mathematical ability. After leaving the Polytechnic School, Fresnel qualified at the department of *ponts et chaussees* as an *ingenieur ordinaire*. For the next eight or so years, Fresnel was engaged in building roads, bridges, irrigation drains, and abutments.

¶ Upon the return of Napoleon from Elba in 1815, Fresnel associated himself with the Royalist cause. Hardly robust, Fresnel was not particularly effective in the military. He returned to Nyons, his usual residence, almost dying. There he was treated with insults, culminating in being dismissed from his engineering duties and being placed under surveillance by the police.

¶ Although cut off from the invigorating intellectual influence of Paris during these years as an engineer, Fresnel gave frequent thought to scientific matters. In 1814, he submitted a paper to the Academie regarding stellar aberration. Unfortunately, by

Augustin Jean Fresnel, French physicist famed for his optical research. *(Bettmann Archive, Inc.)*

being isolated from the centers of research and their attendant libraries, Fresnel was unaware of the contributions of James Bradley (1693 to 1762) and others on this subject. Fresnel's embarrassment was such that he vowed never to submit another memoir without first gaining support from a friend that he had not "broken through open doors."

¶ On December 28, 1814, he wrote to D. F. J. Arago (1786 to 1853), "I do not know what is meant by the polarization of light," and requested that the best works on the subject be sent to him. Eight months later the results of his skillful researches made other savants take notice of him.

¶ Fresnel's investigations began with studies of diffraction. He was able to show that these phenomena entirely correspond to a wave motion if we take account that waves also exhibit interference. That diffraction permits the possibility of propagation in curved lines had already been expressed in Huygen's principle. Fresnel's contribution was the demonstration that diffraction effects are in perfect agreement with this principle and with Young's discovery of interference.

¶ Fresnel studied Young's work, particularly his explanation of the formation of Newton's rings. Young had interpreted the diminution in the size of the rings when the space between the glasses was filled with water as attributable to a decrease in the velocity of light in water which would cause the decrease in wavelength. Moreover, Young had explained the black spot in the middle of Newton's rings by means of an acoustical analogy. Fresnel extended this concept by obtaining interference fringes between two "flat" glass plates wedged at a small angle. He was also able to employ a mirror arranged so as to produce a virtual image of a slit which, with the real slit, simulated Young's "double-slit" experiment.

¶ Together with Arago, Fresnel repeated these experiments with polarized light. These experiments demonstrated that light is, indeed, a transverse wave rather than a longitudinal wave as had previously been thought. How it would be possible for the ether to behave in this way was an enigma, but the observations clearly demonstrated that light must consist of a transverse wave motion.

¶ Fresnel's fame rose. He was elected a member of the Academie des Sciences in 1823, and a foreign member of the Royal Society in 1825. But Fresnel's contributions extended beyond his contributions regarding diffraction.

¶ Arago waxed philosophic in his *eloge* to his friend, pointing out that in the progress of science the question *cui bono?* is frequently asked. Arago noted that, as scientists, one comprehends that the cultivation of the intellect should be the sole occupation of man. But the layman is desirous of finding from science improvements in harvest, in weaving with greater economy, of constructing buildings with greater elegance and solidity, of mining metals and energy supplies, and of eradicating disease. So, the layman asks of science, *cui bono?* Arago then triumphantly described Fresnel's contributions to the construction of lighthouses, which at the time were necessary aids to navigation at sea.

¶ Unaided lamps, of course, are insufficiently bright to be seen over appreciable distances. Reflectors help, but the poor reflectances then available prevented these from being satisfactory in achieving the illuminance levels sought. A single lens having the short focal length required, combined with the requisite diameter, is too heavy to be revolved many times before gears are worn so as to be useless. Fresnel overcame these obstacles by replacing such a lens with one constructed by a process he called *lentilles a echelons* (lenses by steps). In this process, annuli are constructed with radii of different lengths to overcome spherical aberration, and with but sufficient thickness to refract the light appropriately.

¶ Arago calculated that the light produced in this fashion had an illuminating power equal to one-third that which would be produced by uniting all of the gas lights used to illuminate the streets, the shops, and the theaters of Paris (in 1823). Arago added, proudly, that, "France [now] possesses ... the most beautiful light-houses in the world."

¶ Today, we might call Fresnel a "workaholic." He maintained duties as both an engineer and an academician. In addition he assiduously continued scientific research, much of which was undertaken at personal expense. To pay for the required instruments, he became an examiner of pupils at the Polytechnic School. At the close of examinations of 1824, Fresnel suffered an attack of hemoptysis (the spitting up of blood) which forced him to be convalescent. Three years of this exhausted the already debilitated engineer. Arago had the honor of presenting to him the Rumford Medal, on behalf of the Royal Society. Fresnel, with his powers practically dissipated, scarcely glanced at the medal. Then, in a feeble voice, he spoke to Arago, "I thank you for having undertaken this mission. I guess how much it must have cost you, for you have perceived, is it not so? that the most beautiful crown is worth little when it is only deposited on the tomb of a friend!"

¶ Eight days later (July 14, 1827) Fresnel died.   ☺

Brewster's Optical Toy

DAVID BREWSTER (1781 to 1868) is known to every student of optics for his contributions to our knowledge of polarized light, particularly through the study of Brewster's angle. His publications include over 300 technical papers, several books, and numerous reports in encyclopedias, etc. He helped found the British Association for the Advancement of Science and was a fellow of the Royal Society from which he was awarded the Copley medal, the Rumford medal, and six other Royal medals. He was a corresponding member of the French Institute, from which he was honored with prizes for his contributions to optics, and was also president of the Royal Society of Edinburgh. Beyond this, he also achieved many other distinctions. Yet, Brewster's daughter, two years after his death, regarded that his invention of the kaleidoscope, "though of little practical advantage, spread his name far and near, from schoolboy to statesman, from peasant to philosopher, more surely and lastingly than his many noble and useful inventions."

¶ Did you know that Brewster had invented the kaleidoscope? Let us consider how it happened.

¶ David Brewster was born in Jedburgh, Scotland, on December 11, 1781, the third child of the rector of the grammar school in that city. At an early age David was influenced by James Veitch, a craftsman of that area, who encouraged the boy to use his hands to supplement his mind. By the time he was ten, David had constructed a telescope, and the fascination of viewing the stars through this instrument was certainly an inducement toward a life of science.

¶ At the age of twelve, Brewster enrolled at the University of Edinburgh. (It is of interest to note that often when he returned home, Brewster walked the forty-five mile distance!) By his nineteenth year (1800) Brewster received an M.A., and made his first discovery in optics. His career then vacillated between science and the church. He occasionally preached, but his primary support came as a private tutor. This enabled him to pursue his literary and scientific interests. Within a dozen years, Brewster was regularly submitting the results of his research to learned journals.

¶ In 1814, he undertook a series of experiments on the polarization of light by successive reflections between plates of glass. This work was published in the *Philosophical Transactions* for 1815, which won for Brewster the Royal Society's Copley Medal. In this experimental work, Brewster noted "the circular arrangement of images of a candle round a centre, and the multiplication of the sectors formed by the extremities of the plates of glass" used as reflector plates. However, he scarcely gave further attention to the subject at that time.

¶ A year later (1815), Brewster discovered the development of complementary colors by successive reflections of polarized light. Again, he noted the multiple images, which this time were colored. Brewster wrote to J. B. Biot (1774 to 1862) in France that by holding two mirrors "inclined at a very small angle, the two series of reflected images appeared at once in the form of two curves; and that the succession of splendid colours formed a phenomenon which I had no doubt would be considered, by every person who saw it to advantage, as one of the most beautiful in optics."

¶ So, we see that in the investigations of scientific truth, Brewster, in common with many scientists of all time, was distracted by the beauty displayed. And thus the germ of the idea of the kaleidoscope became implanted in his mind.

¶ In subsequent experiments on the action of fluids on polarized lights, Brewster utilized a triangular trough, formed by cementing two plates of glass together with the ends closed up with plate glass. While using these troughs, he noted again a multiplicity of reflections, this time with greater symmetry than noted before.

¶ Brewster next constructed an instrument which attained perfect symmetry, and he eagerly displayed it to entertain some of his colleagues at the Royal Society of Edinburgh. He later conceived the notion that the effect would be enhanced if the object being viewed could move, and subsequently he used a convex lens to image distant objects onto the end of the reflector. In this form, the kaleidoscope was completed.

¶ Brewster felt that in addition to being a toy of some interest, this instrument might "prove of highest service in all the ornamental arts." Accordingly, he sought to patent the kaleidoscope and permit it to be manufactured. Unfortunately for Brewster, he had prematurely exhibited the instrument to London opticians before he had obtained a patent, and this permitted his idea to be pirated.

¶ In the first three months of its being available, some 200,000 kaleidoscopes were sold in London and Paris. Brewster's daughter noted, "This beautiful little toy, with its marvelous witcheries of light and colour, spread over Europe and America with a *furor* which is now scarcely credible." As a consequence, she complained, "thousands of pounds of profit went into other pockets than those of the inventor, who never realized a farthing by it."

¶ I suspect that Brewster himself, even though a Scotsman, was not particularly perturbed at foregoing these profits. He seems to have been more concerned with sharing the knowledge of this optical device and the beauty and entertainment to be derived from it. And certainly we are struck more forcibly by Brewster's substantial contributions than by his invention of this toy.   ☼

Letters to a German Princess

Leonhard Euler

RITING TECHNICAL MATERIAL in an informative manner for the layman is an art that few of us possess. It requires both a thorough understanding of science and an ability to express it in terms that are at once accurate and interesting. Many of the so-called popularizers of science today fail, in my opinion, because they sacrifice accuracy to provide an exotic impact. One of the early attempts to present science in meaningful terms for the layman (and possibly still one of the best ever) was Leonhard Euler (1707 to 1783). His *Letters to a German Princess* are masterpieces of scientific exposition.

¶ Euler was born in Basle on April 15, 1707, to Paul Euler and Margaret Brucker. The elder Euler at that time was a pastor in a nearby village. However, he had studied mathematics under the celebrated Jacques Bernoulli (1654 to 1705), and he instilled this love for mathematics in his son. Leonhard eventually entered the University of Basle where he studied under Jean Bernoulli (1667 to 1748), the younger brother of Jacques. While there, Euler befriended Daniel and Nicholas, the two sons of Jean. At the age of sixteen, Euler received an M.A. degree at Basle.

¶ Meanwhile, Daniel and Nicholas had accepted an invitation of Empress Catherine I to become members of the Academy of Sciences at St. Petersburg, and they used their influence to procure an appointment there for Euler. As he left for Russia, Euler was saddened to learn of the death of Nicholas, and on the day he arrived in that country, the Empress died. This threatened to dissolve the Academy, of which she had laid the foundation. Euler resolved to join the Russian navy. Fortunately, a change in public affairs in 1730 permitted Euler to obtain a situation in Russia as Professor of Natural Philosophy.

¶ At the age of twenty-eight, Euler undertook the solution of an intricate mathematical problem proposed by the Russian Academy. He solved the problem in three days of intensive work, but the exertion of his mind was so great that he lost the sight of one eye, and nearly lost his life.

¶ In June 1741, Euler accepted an invitation from the King of Prussia to continue his research and teaching in Berlin. While there he was introduced to the Princess Anhalt Dessau, the niece of the King. She was desirous of receiving instructions in the various branches of natural philosophy. The *Letters* resulted from this association.

¶ In 1760 the Russian army, while invading Germany, pillaged a farm possessed by Euler. Upon learning that it was Euler's property, the Russian general repaid the loss with a substantial sum. The Empress Elizabeth added a present of additional money to this indemnity. Moreover, the Russians had continued to pay Euler a pension originally granted in 1742. This was a powerful inducement to Euler to return and he sought permission from the King of Prussia to do so. It was there that his letters were printed, although they had been written in the period 1760 to 1762 while he still lived in Berlin. They were translated into most of the languages of Europe; those in my possession were translated by Henry Hunter.

¶ Euler's first letter dealt with the magnitude of things in which he attempts an explanation of relative distances. After reminding the Princess of the concept of a foot and of a mile, he states: "Of all the heavenly bodies the *Moon* is nearest to us, being distant only about 30 diameters of the earth, which amounts to 240,000 miles, or 1,238,400,000 feet; but the first computation of 30 diameters of the earth is the clearest idea. The *Sun* is about 400 times farther from us than the moon, and when we say his distance is 12,000 diameters of the earth, we have a much clearer idea than if it were expressed in miles or in feet."

¶ Euler then discussed velocity, sound (including the pleasures to be derived therefrom), the characteristics of air, and the reason why the air is cold at mountain levels. On June 7, 1760, Euler wrote his seventeenth letter, which was the first dealing with optics. In it, he begins, "Having spoken of the rays of the sun which are the focus of all the heat and light that we enjoy, you will undoubtedly ask, What are these rays? This is, beyond question, one of the most important inquiries in physics, as from it an infinite number of phenomena is derived."

¶ In 1760, the nature of light was very much in question. Euler presented both the corpuscular and wave theories of light, then proceeded to examine the consequences.

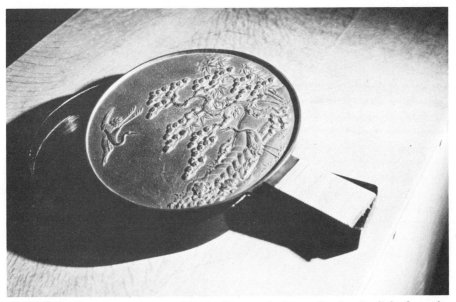

The "Magic Mirror" was invented in the East, and when reflecting Sunlight from the plane surface (opposite that shown) would soon display an image of the back surface (shown). This was caused by the uneven expansion of the metal surface when heated by the sunlight. The mirror is located at the Rijksmuseum voor de Geschiedenis der Natuurwetenschappen, Leyden, Netherlands, where it was photographed by the author.

He pointed out that if one accepted the Newtonian view, small portions of the sun must be emitted in the form of light. On the other hand, Descartes's view that light is a wave motion analagous to that of sound would require an ether. Euler rejected the Newtonian concept on the basis that the sun has not appreciably diminished even though the light particles are thought to be inconceivably small. He concluded his first letter on light, "You will certainly be astonished that it could have been conceived by so great a man, and embraced by so many enlightened philosophers. But it is long since that Cicero remarked, that nothing so absurd can be imagined as to find no supporter among philosophers. For my own part, I am too little a philosopher to adopt the opinion in question."

¶ Having indicated the necessity of an ether to support the light wave motion, Euler described the properties it must have. To permit matter to travel freely through it, it must be extremely tenuous, and this would account for the high velocity of light. Euler clearly and logically accounts for his conclusions. Although subsequent observations have contradicted some of his assertions, his presentations are delightful to read. Moreover, much of his perception was in advance of his time.

¶ Following his description of the nature of light, Euler wrote that Newton attributed color to the nature of the rays. Others, Euler chided, pretend that colors "exist only in ourselves." This, he commented, "is an admirable way to conceal ignorance." Euler attributed color, correctly, as determined by the frequency of vibration of light. (This was nearly forty years before Thomas Young would voice the same opinion.)

¶ In some of the fascinating letters which follow his explanation of color, Euler discusses transparency and refraction. With this background, Euler described the "azure" of the sky to "the want of transparency in the air. The sky is loaded with a great quantity of small particles, which are not perfectly transparent, but which, being illuminated by the rays of the sun, receive from them a motion of vibration, which produce new rays proper to these particles." Although this strikes one as a modern concept of scattering, Euler apparently was simply considering the particles to be blue in color. This would not only explain the blue of the sky but would also account, as he suggested, for the bluish cast of a mountain as seen from a distance.

¶ Euler wrote next of the role of psychology in viewing objects, noting "that we frequently imagine that we can determine by the eye the magnitude and distance of an object; this is not an act of vision, but of understanding." He also noted that the moon appears to be larger when near the horizon than when overhead. He wrote to the Princess that this is "another phenomenon well known to everyone, and which has given occasion to many disputes among the learned." Euler reasoned that this apparent size difference is attributable to an ability to compare the apparent size with known objects on the horizon which are simultaneously sighted.

¶ The Princess was next told of the properties of mirrors and lenses. Four letters were then devoted to descriptions of the eye. These letters indicated, in lucid style, a knowledge of the anatomy of the eye as understood at the time by but a few.

¶ Euler's letters then went on to discuss such topics as gravity, the essence of bodies, inertia, the soul, morals, the state of the soul after death, language, syllogisms, evil and sin, true happiness, human knowledge, truth, monads, and other topics, before returning again to optics. While all of these are interesting, I shall be content to consider only the optics letters here.

¶ In June 1761, Euler returned to the nature of colors, again forming the analogy of colors and sounds. And again Euler emphasized that color is attributable to the frequency of vibration of the ether. He also attributed the color of solid bodies to their ability to vibrate in resonance with the frequency of the light impinging upon them.

¶ Euler's letters then considered electricity, magnetism, and the determination of latitude and longitude. Returning to optics, he described the characteristics of lenses and instruments formed by a combination of lenses. Euler was aware of both spherical and chromatic aberrations, and appeared near to describing a way to correct these defects. The letters conclude with some considerations of astronomical observations.

¶ Shortly after returning to St. Petersburg, Euler lost the sight of his other eye. Blindness did not deter him from his remarkable mathematical researches. His children and his students helped write his memoirs. A servant, ignorant of mathematics, took dictation and prepared manuscripts. In the course of seven years, Euler contributed seventy memoirs to the Academy at St. Petersburg.

¶ In 1771, a widespread fire in St. Petersburg reached Euler's residence. Fortunately, a native of Basle courageously crossed the flames and brought Euler out on his shoulders. Although Euler's library was consumed his manuscripts were saved.

¶ On September 7, 1783, Euler dined with a relative, discussed some of the matters then being considered by philosophers, such as Herschel's discovery of a new planet, and then while amusing himself with a grandchild, he had a stroke and died.

¶ We remember Euler today as the most prolific mathematician of all time. His contributions to optics were minimal, but his letters to a German Princess represent a superb example of presenting an exposition of optical phenomena to the layman. His penetrating style must certainly have been appreciated by the Princess, and I feel that modern authors would do well to approach it.   ☙

Extending the Optical Spectrum

Macedonio Melloni

WILLIAM HERSCHEL (1738 to 1822) had shown in 1800 that optical radiation extends beyond the visible red rays, but his thermometers were too insensitive to detect energy beyond that quite close to the visible spectrum. Another quarter of a century elapsed before technology advanced sufficiently for such measurements to be made. The measurements that were then made were beset with obstacles arising from political upheavals and cholera epidemics as well as scientific uncertainties. These probably would have deterred most of us, but one man of sterner stuff undertook pioneering research that opened a new vista in optical phenomenology.

¶ Macedonio Melloni was born in Parma, Italy, on April 11, 1798. His father was an affluent and well-known merchant; his mother was the daughter of a French physician. Together they provided the boy with an atmosphere of learning and culture. This was fortunate because at the time, the French Revolution had taken place and the Napoleonic influence was being felt throughout Europe.

¶ As a youngster, Melloni would rise early and climb to the top of a hill to see the sun rise. He wrote, "My mind would be completely absorbed by the awakening of nature (as the sun rose). The effect of light on the animals was plain to see, but how does this come about? What is light and how does it come to earth?"

¶ Thus arose Melloni's passion for physics. But he also studied music and art, receiving prizes for his paintings from the University of Parma. In 1819, he completed his B.A. degree at Parma, whereupon his father took him to Paris in hopes that the young Melloni would study the art of engraving at the famous Bervich school. However, Macedonio told his father that he would prefer to study physical and mathematical sciences since he enjoyed these pursuits. His father did not oppose this, and indeed helped his son as much as he could to make use of all the facilities that Paris had to offer.

¶ Melloni stayed in Paris for five years. This must have been a rewarding experience, for the liberal ideas of the time particularly influenced academicians. Although Napoleon had been exiled to St. Helena in 1815, Louis XVIII, who reigned from 1814 to 1824, sought to preserve the reforms achieved by the revolution.

¶ Returning to Parma in 1824, Melloni received an appointment as honorary substitute professor of theoretical and practical physics at the University of Parma. Three years later, he was promoted to full professor and director of the Institute of Physics at a salary of 950 lire as a professor and 250 lire as director. This salary was less than that paid to a palace maid.

¶ In conjunction with Leopoldo Nobili (1784 to 1835), a Professor of Physics at the nearby Florence Museum and a pioneer in the field of electrochemistry, Melloni devised a "thermo-multiplicateur" (i.e., thermopile), consisting of thirty-eight elements of antimony and bismuth. By means of this detector, Melloni was able to measure temperature variations undetectable by any other instrument. This enabled him to begin an outstanding career in research into the characteristics of radiant heat. Thus, our understanding of the infrared spectral region began.

¶ In 1832, when Nobili and Melloni had developed the thermo-multiplicateur, the wave theory of light was still not indisputable. However, Melloni advocated such a theory and asserted that the undulations of radiant heat would be longer than those of visible light if the calorific source is dark, but they would be equal if the source also emitted light. Melloni also noted that thermal radiation from nonluminous bodies was not transmitted through water. He argued that, since the eye contains water, that is the reason we are not visually sensitive to thermal radiation.

¶ In these experiments, Melloni was literally groping in the dark. No standard sources existed. Knowledge of the spectral transmission of materials was unknown. There was no way to "label" a ray. A commonly used source was the so-called Leslie cube, which consisted of a hollow cube filled with hot water. The four perpendicular surfaces were finished dissimilarly to furnish four varieties of radiation characteristics due to their different emissivities. Another popular source was a spiral of platinum wire mounted over the wick of an alcohol lamp.

¶ Melloni used a galvanometer to measure the feeble voltage emitted by his detector. He realized that the output signal was proportional to the radiant energy of the input radiation, but that the deflection of the galvanometer was not linearly related to that signal. He compensated for this by placing a shunt across the galvanometer to reduce the current by a factor of four. He considered that thereby he was operating within the linear range.

¶ Melloni also recognized that measurements based on the signal level reached are not as significant as are those based upon null signal derivations. To obtain a null voltage, Melloni exposed one junction of his thermopile to the unknown source and the other end to a reference, or managed, source. In this manner, the level of the unknown radiation could be compared to a "standard."

¶ These researches were done under trying circumstances, for the political scene in Europe was far from stable at that time. Although Louis XVIII was a liberal king and had introduced periodic elections, freedom of the press, and some other democratic concepts, it was only the beginning of a truly liberal government. The Ultraroyalists meanwhile longed to restore the former privileges of the old aristocracy. They were supported by the aristocratic governments of Europe, forming the Holy Alliance. When Louis died in 1824, his brother, Charles X, was crowned King of France. He was an Ultraroyalist, so many of the liberal concepts which had been introduced were abandoned. Moreover, Charles appointed the most bigoted conservative and churchman among the French nobility as his confidant and representative. He succeeded in antagonizing the opposition beyond anything previously done, resulting in a meeting of the Assemblies in March 1830 to demand the dismissal of the actively conservative ministers. Charles answered by dissolving the Assemblies and in essence restoring arbitrary government. Trouble erupted in Paris. Three days of fighting (the July Revolution) enabled the liberals to gain possession of all of Paris. Charles was forced to abdicate.

¶ Melloni delivered his first lecture at Parma on November 15, 1830. He had warm

A Melloni Bench, such as this, is patterned after that developed by Macedonio Melloni, and consists of a source, some means of affecting the radiation, and a "thermo-multi-plicateur." The source shown here is a spiral of platinum, heated by an alcohol flame. Photographed by the author at Teylers Museum, Haarlem, Netherlands.

words for his country, and described to the students how the Liberals in Paris had struck down the barricades in the July Revolution, stating, "Inspire yourselves with this love of country, this contempt of life, this glorious example." His lecture was received with much applause. The next day he was dismissed from his position, and exiled! He returned to Paris.

¶ The revolutionary movement, however, spread to Parma early the following year. Melloni was called back and made part of the provisional government. It didn't last long. In March the Austrians invaded Italy and Melloni was forced to flee to Paris once again.

¶ Fortunately, Melloni had befriended D. F. J. Arago (1786 to 1853) during his student days. Arago was influential throughout Europe and enabled Melloni to obtain a position as Professor of Physics in the small provincial town of Dole, France. This was not an invigorating scientific environment, so Melloni soon left for Geneva, where he stayed a short period, doing research on thermal radiation. He then returned to Paris where the French government provided him with a modest stipend to continue his research.

¶ From 1832 to 1836 Melloni did his major research on thermal radiation, despite his frequent moves. In 1835, the Royal Society awarded him the Rumford Medal. Numerous other awards followed. In all, Melloni published some 112 papers and memoirs.

¶ Melloni longed to return to his homeland. He couldn't hope to regain his professorship at Parma, as that had been given to another (who, incidentally, contributed nothing at all to science). Again Arago helped his friend. The King of Naples was urged to accolade Melloni with a position in Italy, as he was recognized as one of the foremost of scientists whose experiments formed a new branch of physics. In 1839, the King nominated Melloni as Director of the Conservatory of Arts and Sciences and of the Institute of Meteorology.

¶ From 1841 to 1847 a Meteorological Observatory was under construction on Vesuvius. The prospects for Melloni appeared bright. Then, in 1848 new fighting erupted in Italy and the liberal Melloni was removed as director of the observatory. Indeed, even the observatory was abandoned. Melloni, at the age of fifty-one, retired to Villa Moretta di Portici, near Naples. Six years later, he contracted cholera and died on August 12, 1854.

¶ Although Melloni was considered the "Newton of the theory of warmth," he is not well known today. Possibly, this is because almost all of his manuscripts and perhaps also the second part of his outstanding treatise on thermal radiation, *Termocrisi*, were destroyed in fear that they were contaminated by the cholera that was raging in Naples at the time of his death.

¶ Certainly, Melloni had a turbulent life. The political upheavals of his days affected his ability to create, but his researches have done much to introduce us to infrared radiation. We remember Melloni as a great scientist and as one who believed that in addition to scientific interest there is room for the interests of one's country which "a man of morals may not neglect."

New Insights into Wavelengths

HEN THOMAS YOUNG (1773 to 1829) discovered interference as a result of his "double-slit" experiment early in the nineteenth century, he could obtain no better than an estimate of the wavelengths of light. He had no "benchmarks" to guide him and could only conclude, "the undulations constituting the extreme red light must be supposed to be, in air, about one 36-thousandths of an inch, and those of the extreme violet about one 60-thousandths." While these values are quite accurate, they permit but limited quantitative calculations of optical phenomena. Within about twenty years, however, means were found to provide the benchmarks necessary for quantitative determinations.

¶ Josef von Fraunhofer was born on March 6, 1787, at Straubing, a Bavarian village on the Danube River. He was the tenth child of a master glassmaker who lived in poor circumstances. At an early age, Josef was orphaned and at twelve he was apprenticed to a mirror maker in Munich, where he also had to serve in the house and kitchen.

¶ Quite miraculously, Fraunhofer survived the collapse of his master's house. Although buried in the ruins, he was extricated unharmed. The ruling Prince of Bavaria was touched by this event and bestowed upon young Fraunhofer a substantial sum of money plus some books. The boy used part of this money to buy his freedom and part to buy a glass grinding machine. The teenaged Fraunhofer considered himself an entrepreneur. But his venture was unsuccessful and he returned to his former employer.

¶ Five years later, Fraunhofer joined the Mechanical-Optical Institute of J. von Utzschneider in the Bavarian village of Benediktbeuern. Fraunhofer at this time (1806) was just nineteen. However, he made dramatic improvements in the grinding and polishing machines. He also became interested in the manufacture of glass. His efforts in these areas raised the company to one of fame.

¶ Glass manufactured at that time often contained numerous striae, which prevented one from making optical components of high quality. Moreover, to make an achromatic lens one must know both the refractive index and the dispersion. Since these quantities are best expressed as a function of wavelength, it was not possible to provide quantitative evaluations at that time.

¶ Fraunhofer's efforts led to the manufacture of large pieces of flint glass free of striae, and also in other respects suitable for the finest optical purposes. He also introduced an exact control over the refractive index of the glass by means of prisms cut from the glasses to be tested. With these excellent prisms, he obtained spectra of greater purity than had Newton. He fully recognized the need to use a fine slit to admit light onto the prism and the requirement that the rays be parallel upon striking the prism. Moreover, he found that examination of the spectrum through a telescope afforded the greatest accuracy in the examination of the spectrum. In essence, the modern spectroscope was born. Of greater importance, undoubtedly, was the discovery of dark lines in the solar spectrum!

¶ Fraunhofer proved by exact measurements of the position of these lines, with many

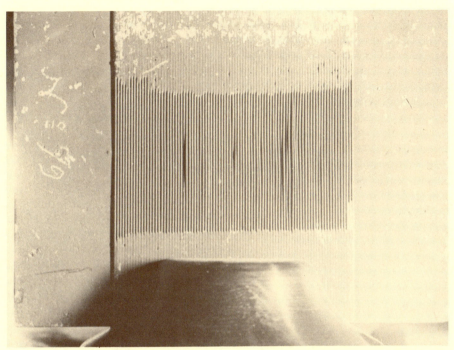

This grating, approximately 17 by 28.5 mm in size, was made by Fraunhofer in his investigations of the diffraction of light. It contains 87 wires, approximately 3 lines per millimeter. *(Photographed by the author at the Deutsches Museum in Munich)*

Fraunhofer's glassworks in 1820. *(Photo courtesy of Deutsches Museum, Munich)*

variations in the experimental conditions, that "these lines lie in the nature of the sun's light, and do not arise by diffraction, self-deception, etc." Fraunhofer also found the same lines in the spectrum of Venus. However, the spectra of the brighter stars which he observed displayed lines quite differently grouped. Fraunhofer modestly refrained from conjecturing on the cause of these lines, either in the solar spectrum or in those of the stars.

¶ By providing an exact drawing of the solar spectrum he identified several of the lines he observed, using the letters from A in the red to H in the violet. This nomenclature is still used today. Now, Fraunhofer could obtain the benchmark necessary to select a specific kind of light.

¶ With glass of known refractive index and dispersion, and also free of striae, construction of larger telescopes was now feasible. By means of a Fraunhofer telescope supplied to the observatory in Konigsberg, it was possible for the first time to discern parallax in a fixed star. This had been sought since the time of Copernicus and had occupied the efforts of numerous astronomers. Star No. 61 in the constellation Cygni showed a displacement every half year amounting to 0.3 seconds of arc, from which it can be calculated that it is ten light years distant. The second star for which a parallax was found, Vega, was also measured with a Fraunhofer telescope. This was the "giant" Dorpat refractor of four meters in length and having an objective aperture of 25 centimeters.

¶ Fraunhofer later became interested in diffraction. During the period that Fresnel was occupied with his investigations of diffraction which led to the establishment of the wave theory of light, Fraunhofer developed diffraction gratings. His first grating consisted of fine wires, stretched exactly parallel to one another and at exactly equal distances apart. Later gratings which he devised used wires from 0.04 to 0.6 millimeters in diameter. Such a device has the important advantage of greater dispersion from which wavelengths can be precisely determined. For instance, Fraunhofer found that the D lines of sodium (as they are now known to be) are between 0.0005882 and 0.0005897 millimeters in length.

¶ Not content with gratings made in this fashion, Fraunhofer coated plain glass with gold leaf and then engraved on it very fine lines with a diamond point controlled by a dividing engine. By this means he was able to construct gratings with 300 lines to the millimeter. Fraunhofer published his results, but this had little immediate impact on the scientific community. After all, Fraunhofer was a technician with no formal training.

¶ Fraunhofer lived a simple and modest life. He never married. He scarcely allowed himself any rest from his activities in the optical factory. In 1825 he fell ill with inflammation of the lungs, but still continued to work. He never recovered from that disease, and died on June 7, 1826, at the age of thirty-nine.

¶ Although most of his contemporaries did not bring him acclaim, the Berlin Academy did pay him honor, and his grave in Munich bears the epitaph *Approximavit sidera* (He brought the stars nearer). His efforts have made a lasting impression on science, particularly in our understanding of the nature of light and its application to spectroscopy and astrophysics.

The Evolution of Spectroscopy

WHEN JOSEF VON FRAUNHOFER (1787 to 1826) discovered dark lines in the solar spectrum in 1814, he opened an entirely new field of optical investigations. Appropriately, it began with Fraunhofer's continued investigations of the spectral characteristics of the light emitted by various lamps. The paths to the use of spectroscopy as an investigative tool, however, had several bridges yet to cross. It is of interest to review some of these.

¶ In 1822, David Brewster (1781 to 1868) undertook a series of investigations to discern by "chemical analysis" the absorption of sunlight by transparent colored media. To do this, he passed sunlight first through different colored materials and then through a prism. He found, for example, that the blue glass that he used transmitted the extreme red rays as well as the blue and violet. In that same year, John Herschel (1792 to 1871) investigated the spectra of a number of luminous flames. Among his findings were that sulphur, while burning feebly, emits primarily blue and violet rays, but when buring vigorously it emits a homogeneous and brilliant light.

¶ William Henry Fox Talbot (1800 to 1877), the inventor of the negative/positive photographic process, examined the spectra of spirit lamps impregnated with common salt. He succeeded in obtaining a bright source, but in subsequent experiments he failed to attribute the yellow rays to any specific chemical despite attempts to steep the wick in the muriate, sulphate, or carbonate of *soda*. Talbot did find that candles and platinum gave a strong yellow light when they had previously been touched by hand. Common salt sprinkled on platinum also gave rise to this light, as did wetting the platinum. This led Talbot to conclude that the yellow light was related to the water rather than to the sodium. This conclusion was reinforced by noting that potassium salts did not produce the yellow light. However, when he discovered that the light was also produced with sulphur, he had to abandon this notion.

¶ The ubiquitous yellow line thus perplexed spectroscopists and constituted a major obstacle in the understanding of the origin of this radiation. In 1834, Talbot did investigate lithium and strontium flames, noting that in the spectrum of the latter element "a great number of red rays well separated from each other by dark intervals" exist, whereas in lithium he observed but one red ray. He concluded, "Hence, I hesitate not to say that optical analysis can distinguish the minutest portions of those two substances from each other with as much certainty, if not more, than any other known method."

¶ In the following year, Talbot proposed that an extensive course of experiments on the spectra of chemical flames be undertaken. Although he did not heed his own advice, Brewster apparently did and in 1842 reported that he had investigated "200 or 300" minerals, salts, and other substances. His results were inconclusive, although he did mention that "Fraunhofer's luminous D-line" was present in nearly every spectrum. He did not report any extensions of this work.

¶ In addition to the emission lines, observations of dark lines were also reported. Talbot, for instance, while investigating the spectra of copper salts, boric acid, and barium nitrate found dark lines similar to those of the solar spectrum. Very likely, Talbot was viewing the spectra of undissociated compounds. In 1838, John William Draper (1811 to 1882), the University of the City of New York chemist, demonstrated that the dark lines were spurious and attributable to the presence of some incombustible substance in the flame. He also found several lines in addition to the yellow line in the spectra of

To obtain higher resolving power, spectroscopists began to "gang" several prisms together. Here five dispersing prisms are used with a sixth arranged to permit the light to traverse through the prisms twice. *(Photographed by the author at Hilger-Watts in London)*

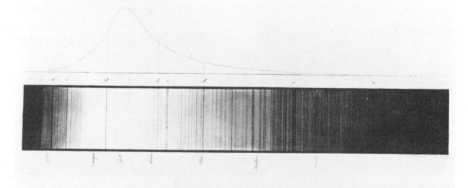

Spectral analysis evolved by attempts to unravel the cause of the dark line in the solar spectrum, first observed by Fraunhofer. *(Photographed from an old textbook)*

the monochromatic sources others had investigated.

¶ Chemical analysis by the utilization of spectra had reached an impasse. The complexity of the observed spectra, plus the ubiquity of the D-line, undoubtedly formed the barrier toward a comprehension of the responsible mechanisms. This was resolved in the 1850s by the collaboration of two rare men of science, Robert Wilhelm Bunsen (1811 to 1899) and Gustav Robert Kirchhoff (1824 to 1887).

¶ Kirchhoff was born at Konigsberg, Prussia, on March 12, 1824. After completing his formal education there, he became a *Privat-Dozent* at Berlin. in 1850 he was appointed Extraordinary Professor of Physics at Breslau, where he met Bunsen. Four years later, he joined Bunsen at Heidelberg.

¶ In 1859, Kirchhoff published a paper in which the Fraunhofer lines were explained as absorption by cooler gases in the solar atmosphere. He then advanced his famous law which states that the ratio of the emissive power of a body at a given wavelength to its absorptive power is a constant at a particular temperature.

¶ Bunsen was born at Gottingen on March 31, 1811. His father was the librarian and Professor of Linguistics at the university there. Bunsen received his Ph.D. at that university in 1830 and after other appointments he accepted the Chair of Chemistry at Heidelberg in 1852. Bunsen was an experimentalist, instilling in his students the practice of practical techniques.

¶ Bunsen's ability to experimentally determine the essence of an investigation was essential to an understanding of flame spectra. He seemed to have limitless inventive power to create new means of investigating spectra which, combined with Kirchhoff's insight, provided the results that brought them fame. This is indicated by one of Bunsen's first introductions to their collaboration. Bunsen realized that a clean flame is essential to the investigations planned and that an alcohol lamp with its attendant wick does not meet this requirement. Bunsen's burner was the result, a device still in use.

¶ Observing the spectra of clean flames, Bunsen discovered the spectra of two new elements within about five years of the introduction of his burner. This was done by recognizing that the observed emission lines are unique for each element. Moreover, the absorption lines originate in an analagous fashion. Kirchhoff was thus able to prove the existence of chemicals in the sun which are responsible for the observed Fraunhofer lines.

¶ Spectroscopists, of course, sought higher resolving powers and to this end Kirchhoff placed three prisms in series. Others followed with even greater numbers. The art of ruling gratings improved to such an extent that Henry Augustus Rowland (1848 to 1901) was able to rule over 500 lines to the millimeter on blanks up to fifteen centimeters in width. Rowland, who achieved greater success in engineering feats than in physics, also ruled gratings on concave mirrors which enabled spectra to be photographed without the use of prisms or lenses.

¶ Today, of course, with the usé of lasers and computers, the resolving power attained at the turn of the nineteenth century seems somewhat tame. Reflect on the fact, however, that spectral analysis as introduced by Bunsen and Kirchhoff led to discoveries as revolutionary as those of Newton. It enabled man to discover the chemical composition existing beyond the earth and solar system into the reaches of the cosmos. ⊗

A Glance at Nineteenth-Century Science

URING THE NINETEENTH CENTURY, advances in technology, a changed economic picture, new political structures, and an evolution in the attitude toward morals affected the way science was practiced and evaluated. Since the historian of science places emphasis on scientific accomplishments, these anecdotes must reflect the underlying activity responsible for these gains in optics. A brief overview of some of the general scientific activity will thus enable us to better appreciate how optical advances were achieved during this period.

¶ The study of heat and its transformation was one of great intellectual, as well as technical and economic, importance. The principle of the conservation of energy was possibly the greatest physical discovery of the mid-nineteenth century. It showed that mechanical work, electricity, and heat are different forms of energy, and thus brought together several facets of science. The laws of thermodynamics indicated that not only the quantity of energy, but its availability, are what matter. The knowledge of thermodynamics also began to play a larger part in such fields as chemistry and biology in the nineteenth century. It was also to make its mark on optics.

¶ During the nineteenth century, the continuous interplay between the growing requirements of commerce and industry created a growing emphasis on engineering. New means of operations, which included machines, engines, and materials, had to be considered. For instance, the need for more yarn and more cloth led to the development of textile machinery; the need for more coal led to the development of the steam engine; cheap transportation requirements led to improvements in ports, canals, roads, bridges, and finally to the innovation of the railroad. As these developments took place, they were adapted to previously unthought-of uses. The steam engine, for example, was developed for pumping. It was then adapted to blowing furnaces, and then to supplant the water wheel to drive machinery. Still later, steam engines were mounted on a boat or wagon, becoming a steamship or railway.

¶ Engineers, in large part, began as simple workmen but were more often than not skillful and ambitious. Often they were illiterate or self-taught. As technology placed increased demands on them, they sought scientific enlightenment. Similarly, the academician was forced to leave his sanctuary in order to gain access to the technology required for the more involved experimental procedures. This interplay of science and technology resulted in improvements in both areas. The concepts of thermodynamics aided the development of machines and engines with improved performance characteristics. It also led to revolutionary concepts, such as the turbine, the internal combustion engine, and the refrigerator.

¶ Electricity was hardly more than a laboratory curiosity in the eighteenth century. The discovery by Hans Christian Oersted (1777 to 1851) of the magnetic effect of an electric current led to the electric telegraph and the electric motor. In 1831, Michael Faraday (1791 to 1867) demonstrated the relationship between magnetism and electricity. From this it was possible to generate electric currents by mechanical action. Moreover, Faraday's qualitative intuitions were given a precise and quantitative mathematical form by James Clerk Maxwell (1831 to 1879).

¶ Although perhaps not pertinent to our optical anecdotes, the progress of biology in the nineteenth century awakened man to an understanding of his relationship with other living things. The publication of *On the Origin of Species by Means of Natural Selection* by Charles Robert Darwin (1809 to 1882) in 1859 aroused a long and bitter controversy. But the matter seemed to be more political and theological than scientific, and biologists generally found Darwin's conclusions to have an enormous liberation effect. The book provided a unifying principle to the living world.

¶ In 1835, Louis Pasteur (1822 to 1895) came into contact with the activities of living ferments. His subsequent experiments led him to the conclusion that fermentation is due to living organisms rather than inert chemical reactions. Further research led to the effective foundation of scientific medicine.

¶ Science was in a stage of transition in the nineteenth century. From a benign assembler of observations of nature, science became a material force that changed the pattern of life. Scientists could not avoid considering the consequences of their ideas and the fruits of its handmaiden, technology. Many scientists avoided the unpleasantness associated with this change by seeking refuge in the pursuit of "pure" science. They reasoned that if they received no material profit from their activities they would be free of any blame of the consequence. This attitude tended to color their ideas and even the theories advanced. Science, in this view, was one of blind fate. Fortunately, this attitude did not persist.

¶ Science appeared to be finite. It seemed that the progress attained in joining together the concepts of heat, light, electricity, and magnetism into a single electromagnetic theory signaled a conclusion to our search for the meaning of nature. The view that the future of the universe could be predicted once the motion and position of all particles is known seemed justified.

¶ They were concepts that were rudely dissipated as the nineteenth century closed and the twentieth century began.   ☙

The Exploits of an Optical Observer

Samuel Pierpont Langley

ODAY, OPTICAL ASTRONOMY makes frequent use of satellites and rockets to enable observations to be made from above the turbulent and partially opaque atmosphere. Although some of these observations can be made with computer-controlled instruments, human observers are still frequently required. Astronauts have justifiably become modern heroes as a result of the risks and daring associated with their missions. It would be a mistake, however, to consider that risk and daring are innovations brought to optical observations by the space age. Reflect, for example, on the exploits of Samuel Pierpont Langley (1834 to 1906).

¶ Langley was born on August 22, 1834, at Roxbury, Massachusetts. He was descended from families which had settled in New England early in the seventeenth century, and among his ancestors were skilled artisans, mechanics, clergymen, and a president of Harvard College. (The last was also the author of the first American work on astronomy.) Langley recorded that his father owned a telescope which he used to study the stars as well as to watch the building of the Bunker Hill Monument. Langley also devoured books of all sorts, but concentrated his studies on mathematics. Upon completion of high school, he studied civil engineering and architecture, although he did not enter college.

¶ After working as an engineer in Chicago and St. Louis for seven years, Langley returned to New England and spent some time building telescopes. He then spent a year traveling in Europe visiting scientific institutions, observatories, and art galleries. Upon his return he had yet made no significant contributions to science in the way of publications or speeches; nor had he gained the attention of scientists. Nevertheless, he possessed a quality that inspired respect and won confidence.

¶ He obtained an assistant's place in the Harvard Observatory; two years later he was awarded a professorship of mathematics and direction of the observatory at Annapolis; and at the age of thirty-two he became the Professor of Astronomy at the Western University of Pennsylvania and Director of the Allegheny Observatory.

¶ Upon arriving in Pittsburgh, he found the observatory equipped with an equatorial telescope which had been used by an amateur group for star gazing, but virtually nothing else—no transit, not even a clock. To obtain money to equip this observatory, Langley had the insight and foresight to provide the railways with a needed standard of time. Although this had been done in places such as Washington and Cambridge, Langley introduced it to the West. He established a clock which provided the exact time, twice a day and automatically, to every railway station on lines extending for eight thousand miles.

¶ In about 1873, he began scientific studies of the sun. Since photography was still in its infancy, Langley sketched pictures showing details of the sun's surface, particularly sun spots. In 1875, he began extending this work to measurements of the heat spectra of the sun. These researches began with the use of a thermopile, but such a detector is too large to obtain the spectral resolution he desired. Moreover, Langley considered such a detector as merely an indicator of radiation, rather than as a measurer of radiation. In December of 1879 Langley began a research program to develop a superior detector. This was accomplished within a year.

¶ Langley called his invention a bolometer to indicate that the device, indeed, measures radiation. He described his method of preparation:

I roll steel, platinum, or palladium into sheets of from 1/100 to 1/500 of a millimetre thickness; cut from these sheets strips one millimetre wide and one centimetre long, or less; and unite these strips so that the current from a battery of one or more Daniells' cells passes through them. The strips are in two systems, arranged somewhat like a grating; and the current divides, one half passing through each, each being virtually one arm of a Wheatstone's bridge.

¶ With his bolometer, Langley was able to obtain measurable deflections from temperature variations of "1/10,000 of a degree Centigrade." He felt that the ultimate limit would be a tenth of this. Moreover, he asserted, "the instrument is . . . far more prompt than the thermopile." He repeated a number of the measurements previously made by Macedonio Melloni (1798 to 1854), reporting an accuracy of one percent.

¶ Langley realized that the measurement of light dispersed by a prism is subject to numerous experimental difficulties arising from the nonlinear dispersion of the prism and the unequal spectral transmission of the prism materials. To circumvent these problems, Langley "employed two of the admirable gratings of Mr. Rutherford, one containing 17,296 lines to the inch, or 681 to the millimeter, and the other one-half that number, both ruled on speculum metal."

¶ Langley did not use a collimator, rather he placed an entrance slit five meters from the grating. He described this apparatus:

The rays from the grating fell upon a concave speculum (whose principal focal distance was about one meter), and from this were concentrated upon the mouth of the bolometer, forming a narrow spectrum, which passed down the case of the instrument and fell upon the bolometer thread. As this thread moves along the spectrum parallel to the Fraunhofer lines, its coincidence with one of them is notified by a lowering of its temperature and a deflection of the galvanometer.

¶ Prior research by others had indicated that the selective absorption of the atmospheric gases allowed terrestrial temperatures to be sufficient for human survival (the greenhouse effect). In this regard, Langley asserted that without selective absorption in the atmosphere, "the temperature of the soil, even in the tropics at midday under a vertical sun, would fall to some hundreds of degrees below zero." Langley thus reasoned that terrestrial measurements of solar heating cannot provide an "estimate of the amount of solar heat before absorption (the solar constant)." Further, Langley understood (about a century ago) that "Could we ascend the atmosphere, this heat might be directly measured."

¶ Late in 1880, Langley proposed that observations similar to those he had made at the Allegheny Observatory be repeated at "the base and summit of a lofty mountain."

Fortunately, the generosity of "a citizen of Pittsburgh" placed considerable means to outfit an expedition for this purpose. Langley did not disclose the name of his benefactor, simply stating, "By his own wish his name is not mentioned in this connection, but it is proper to acknowledge . . . the timely and indispensable aid which made this project a reality."

¶ Many possible sites were considered. Some, such as Pike's Peak, were rejected because they "are rarely free from mist and cloud during the summer." With the help of the Geological Survey, the Coast Survey, and the Army, Mount Whitney in California was considered to be the most desirable site. "Its eastern slopes are so precipitous that two stations can be found within 12 miles, visible from each other . . . while its summit is almost perpetually clear during June, July, August, and September."

¶ While Mount Whitney promised to be a superb site from a scientific viewpoint, it (in 1881) was a wild region, remote from any railroad, and it was uncertain that heavy instruments could be transported to the summit. Army Captain O. E. Michaelis and a small group of soldiers were detailed to assist Langley.

¶ It was hoped to reach the scene of operations in July. On July 7, 1881, Captain Michaelis, J. E. Keller of the Allegheny Observatory, W. C. Day of the Johns Hopkins University, and Langley left Pittsburgh by train for San Francisco, accompanied by two and a half tons of equipment. They reached San Francisco fifteen days later. There they were joined by eight enlisted men, they hired a carpenter, and were joined by a volunteer.

¶ Riding day and night, Langley and his party reached the base camp on the evening of July 24. Langley was impressed to be "looking up through desert air, where the shade temperature was over 100°F., to the patches of snow on (the mountains) summits." A week later the wagons containing the instruments arrived. They were unpacked, cleaned, and observations were begun immediately. One of the first obstacles then encountered was the heat. Inside a tent set up to form a dark room for the galvanometer, the heat "rose to a point beyond human endurance. . . . After a day's trial this plan was then abandoned." On August 6 it began to rain and the wind covered the instruments with sand and dust. Next, they felt a slight earthquake shock. But by August 11, "the first complete morning, noon, and evening bolometer observations was obtained."

¶ Also on August 11, a team set out on mules to establish a camp near the summit. Langley waited impatiently to hear word of a successful ascent, which was scheduled to take three days. On the second day, Langley could be restrained no longer and he set out for the summit. He described the ascent:

The trail made sharp turns, plunged down into ravines into which descent on the saddle seemed at first impossible; and wormed its way between boulders, and climbed over rocks and fallen timbers, in such a manner as to give a formidable impression of the dangers our apparatus must have incurred in the ascent, though it had taken a somewhat easier and longer route than ours.

¶ Langley soon experienced the difference in character of solar radiation at high altitude from that in the valley. He described his hands as having "the appearance of as severe burns as though they had been held in an actual fire, and my face was hardly recognizable." Langley, moreover, was ill for over a week, even though he camped some 3,000 feet below the summit. That did not prevent him from scaling the peak where he noted the sky to be of "unaccustomed purity" which was "everything I could have hoped to find."

¶ Observations were made at the base site on only three days (August 11, 12, and 13) and at the mountain camp only three days (September 1, 2, and 3). High winds, variable temperatures, desert sand and dust, as well as logistics prevented the more extensive gathering of data. Yet it was an auspicious beginning. Subsequent determinations of the solar constant have followed this daring and ingenious venture by Langley.

¶ In 1887, Langley was appointed to be the third secretary of the Smithsonian Institution. Here, he concentrated much of his effort to prove the feasibility of flight with heavier-than-air machines. In 1896, he was able to successfully accomplish a sustained flight with a model. Subsequent attempts with manned flights were not successful. In 1903, the *New York Times* harshly criticized Langley for wasting public money (Congress had appropriated a pittance for this effort) on this idle dream. Flying, the *Times* said, was a thousand years off. Nine days later the Wright brothers made their successful flight. Langley died less than three years later.

The Demise of the Ether

Albert Michelson

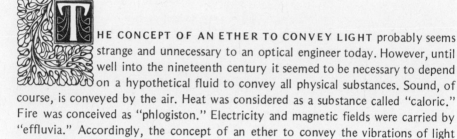 HE CONCEPT OF AN ETHER TO CONVEY LIGHT probably seems strange and unnecessary to an optical engineer today. However, until well into the nineteenth century it seemed to be necessary to depend on a hypothetical fluid to convey all physical substances. Sound, of course, is conveyed by the air. Heat was considered as a substance called "caloric." Fire was conceived as "phlogiston." Electricity and magnetic fields were carried by "effluvia." Accordingly, the concept of an ether to convey the vibrations of light seemed to be quite reasonable.

¶ The concepts of caloric, phlogiston, and effluvia were discarded early in the nineteenth century. And the concept of the ether was questioned. However, the discovery by Christian Doppler (1805 to 1853) in 1842—that the apparent frequency of a wave motion varies in accordance with the velocity of the source relative to an observer—renewed considerations of an ether. The ether must be stationary relative to the earth for the Doppler principle to apply to light waves. George Gabriel Stokes (1819 to 1903) suggested that the earth carries a portion of the ether along, and as a result it is at rest relative to the earth near to the surface. This concept was tested by George Biddell Airy (1801 to 1892), the Astronomer Royal of England, by measuring the astronomical aberration with a telescope filled with water. He found no difference in the aberration, thus indicating that the aberration does not depend upon the thickness of the refracting medium, which is contrary to Stokes's hypothesis of a stagnant ether.

¶ James Clerk Maxwell (1831 to 1879) interpreted light as an electromagnetic wave and incorporated this into a set of partial differential equations. However, Maxwell still considered that light was conducted by a luminiferous ether. It would take more experimental evidence and a great deal of theoretical formulation to rid optics of the need for an ether. That began late in the nineteenth century.

¶ Albert Michelson was born on December 19, 1852, at Strzelno, Poland. His father, Samuel, was a merchant of Jewish descent. Within three years of Albert's birth, they left Poland, presumably to escape the anti-Semitism that arose there. They worked their way across Europe and finally embarked on a steamer bound for New York. There, Samuel heard of the adventures of the "forty-niners" and the sudden fortunes made overnight in the unsettled West. This lured him to California, and he chose to go by way of the Panama Isthmus on ship, train, and mule.

¶ The first stop of this journey was Porto Bello on the Isthmus. Here, they found a slum in which the inhabitants lived in squalor, many ill with malaria, smallpox, or "brain fever." Thieves and robbers looted the dead, the sick, and the unarmed passengers; there were no police. The Michelsons escaped from Porto Bello and traveled by canoe, paddled by natives, through swamps and lakes. They changed to muleback on higher ground, and then took a pioneer railroad to Panama City. Drinking water was exhorbitantly priced; raw sewage flowed in the streets; violence was common.

¶ Several weeks passed before space could be obtained on a ship bound for San Francisco. After another sixty days, the Michelsons reached the Golden Gate. Samuel took his family to Murphy's Camp in the foothills of the Sierra Nevada, about 150 miles east of San Francisco, where het set up a little store.

¶ Albert attended the first public school built in Murphy's Camp. His first-grade teacher there was a fourteen-year-old Irish girl who had been raised in a Spanish-speaking convent, so she spoke but little English. By the time Albert was eight, the American Civil War broke out. Most of the town was sympathetic with the Union cause and when peace was declared, the population was overjoyed. This turned to anguish when the news of Lincoln's assassination was received. Albert received a middle name, Abraham, in compassion for the late president.

¶ In 1866, Michelson began attending the San Francisco Boys's High School, where the principal was sufficiently impressed with the boy's ability that he gave him the job of setting up experiments in the science class. Meanwhile, the gold had been exhausted at Murphy's and the prospectors had moved to Virginia City, Nevada, where silver had been found. Samuel Michelson closed his store in Murphy's and headed for Virginia City. Albert continued his schooling in San Francisco.

¶ Nevada had been admitted to the Union in 1864. Five years later, Samuel noted that a candidate to the Naval Academy would be proposed for the State of Nevada. He encouraged his son Albert to compete for this recommendation, and on June 10, Albert took the examination. Although he tied for first place in the examination, the congressman appointed one of the other boys. Michelson's high school principal and others petitioned the congressman to reconsider. Of course, he could not do that, but he did write to President Grant requesting that Michelson be given a presidential appointment. Michelson then went to Washington for an interview with the President and to solicit that appointment. Grant gently advised the boy that he had already filled his ten available appointments-at-large, but that Albert should go to Annapolis on the chance that a vacancy might occur if one of the appointees failed the exam. Michelson went to Annapolis, but no vacancy developed.

¶ Discouraged, Michelson prepared to return home. As the train was about to leave the station, a messenger summoned Michelson to return to the White House. Grant had been persuaded to make an exception, and Albert received the President's illegal eleventh appointment-at-large.

¶ On May 13, 1873, Albert Michelson graduated ninth in a class of twenty-nine from the Naval Academy. He was at the head of the class in optics and second in heat and climatology (thermodynamics), but his indifference to seamanship (in which he ranked twenty-fifth) lowered his overall status.

¶ After spending a required two years at sea, Michelson returned to Annapolis where he was asked to lecture on the procedure Leon Foucault (1819 to 1868) had used to measure the speed of light. In preparing for this lecture, Michelson was inspired to re-design the experiment. With improved apparatus, he was able to determine the speed of light as 299,910 km/sec, which represented an improvement over what had been reported previously.

¶ In the fall of 1880, Michelson received a year's leave of absence from the Naval Academy and he spent it studying in Europe. His first stop was in Berlin, where he planned to extend his measurements of the velocity of light, this time measuring it in the direction of motion of the earth and simultaneously at right angles. Such measure-

Albert A. Michelson with a group of scientists in January 1931 while attending a meeting at the California Institute of Technology. (Left to right) Walter S. Adams, Albert Michelson, Walter Mayer, Albert Einstein, Max Farrand, and Robert A. Millikan.
From a copy negative of a print probably obtained from the California Institute of Technology for the dedication of Michelson Laboratory on May 8, 1948. The print from which the copy negative was made shows the autographs of A. A. Michelson, Robert A. Millikan, and A. Einstein 1931, which confirm the identities of the three men in the front row and the year. Otherwise, identification is from Bernard Jaffe, *Michelson and the Speed of Light* (1960), facing page 121: "Plate IV. In his last year. A few months before his death in 1931 Michelson posed for this picture with a group of scientists at a Caltech meeting. (Left to right) . . ." And on page 167 he mentions a banquet on 15 January 1931 given by "Caltech's Associates" to honor Dr. and Mrs. Einstein, which Michelson attended. Also, the Movietonews, Inc. library index card covering the recently destroyed footage taken on 7 January 1931 and described as "Pasadena, Calif.—Six Great Intellects Pose for Movietone—Einstein, Dr. Millikan, Mayers, Dr. Adams, Michelson & Farrand . . ." probably refers to the same group and general occasion.

(Photo courtesy United States Naval Academy Museum)

ments would reveal any current of the ether relative to the earth. The instrumentation for this measurement required a beamsplitter. Making a "half-silvered" mirror was pushing the state of the art, and hampered Michelson in reaching his objective.

¶ The following summer, Michelson studied at the University of Heidelberg. There he met Robert Wilhelm Bunsen (1811 to 1899) who had developed the use of spectra for chemical analyses. From him Michelson probably learned the technique of making the needed beamsplitter. He next visited Paris, where he was asked to demonstrate his newly developed interferometer.

¶ In April 1882, Michelson returned to America to accept an appointment with the newly founded Case School of Applied Science in Cleveland. Soon thereafter, he met Edward Williams Morley (1838 to 1923), the Professor of Chemistry there. Together, they set about refining the instrument Michelson had used in Berlin to measure the ether drift. Morley suggested that floating the instrument on mercury would permit the heavy apparatus to be turned steadily on its (vertical) axis. They also modified the former instrument by placing mirrors to reflect the light over a path some ten times longer than that previously used. The results of these observations indicated that no motion of the earth relative to the ether could be detected.

¶ Immediate reaction to this experimental investigation was essentially negligible. Michelson turned his attentions more toward the uses of the interferometer, and in 1889 he was awarded the Rumford Medal by the American Academy of Arts and Sciences for his measurements of the velocity of light and for his interferometer.

¶ During the period that Michelson and Morley were attempting to measure the ether drift, Heinrich Rudolph Hertz (1857 to 1894) was inducing electrical signals over gaps extending to sixty feet. This not only established that electric waves are essentially the same as light waves, it suggested that they were transmitted as etherial waves. Thus, the luminiferous ether became the electromagnetic ether.

¶ Gradually, however, the implications of the Michelson-Morley experiment began to be felt. Suggestions were made that the lengths of matter change with their orientation in the ether field. Additional measurements were performed. No evidence of an ether drift could be found.

¶ In 1905, a then-obscure physicist, Albert Einstein (1879 to 1955) published three papers that significantly changed the concepts of physics. In the third paper he noted, "The introduction of a 'luminiferous ether' will prove to be superfluous." Einstein had eliminated the need for an ether by eliminating the concept of an absolute space.

¶ The Michelson-Morley null results now had theoretical approval. However, when Michelson received the Nobel Prize (the first American scientist to be so honored) no mention was made of the ether-drift experiment. Rather, the award was given in recognition of his original methods for ensuring exactness in measurements, for his investigations in spectroscopy, and for his achievement in obtaining a nonmaterial standard of length. Michelson was also honored by the Royal Society with its Copley Medal, which he cherished as a tribute by other physicists. No mention was made at that award, either, of the ether-drift experiment.

¶ Nevertheless, by virtue of the Michelson-Morley experiment and Einstein's theoretical reasoning, there was no longer a need for an ether. ☙

Quanta and Photons

Max Carl Ernst Ludwig Planck, Albert Einstein

BY THE MID-NINETEENTH CENTURY THE WAVE THEORY of light was well established. One could thus consider relationships existing in a specified spectral region. In fact, the theorem of Gustav Robert Kirchhoff (1824 to 1887) considered the fraction of energy incident in a narrow spectral band on a cavity maintained at a given temperature. He showed that the ratio of the emission to the energy absorbed in that spectral band must be a constant, and that this constant represents an intensity distribution which is a universal function. Such a function is dependent only on the temperature and wavelength, and not on the size, shape, or material of the cavity.

¶ Approximately thirty years later, James Clerk Maxwell (1831 to 1879) showed that electromagnetic waves are governed by differential equations. The consequences of these equations were applied to blackbodies by Ludwig Boltzmann (1844 to 1906) in 1884. In nine more years, Boltzmann's results were decisively extended by Wilhelm Wien (1864 to 1928), resulting in his displacement law. Nevertheless, attempts to mathematically describe the emission characteristics of blackbody radiation eluded the efforts of researchers up to about 1895.

¶ This problem was of interest to Max Carl Ernst Ludwig Planck (1858 to 1947) and he is celebrated for his solution. However, attaining that solution was difficult, for it required a departure from conventional physics.

¶ Planck was born at Kiel, Germany, on April 23, 1858. He was descended from a long line of scholars, lawyers, and public servants. This heritage is evident in his autobiography, which begins:

> *My original decision to devote myself to science was a direct result of the discovery which has never ceased to fill me with enthusiasm since my early youth—the comprehension of the far from obvious fact that the laws of human reasoning coincide with the laws governing the sequences of the impressions we receive from the world about us; that, therefore, pure reasoning can enable man to gain an insight into the mechanisms of the latter. In this connection, it is of paramount importance that the outside world is something independent from man, something absolute, and the quest for the laws which apply to this absolute appeared to me as the most sublime scientific pursuit in life.*

¶ Planck studied at the *Maximilian-Gymnasium* in Munich where his view was bolstered by his introduction to the laws of physics. He never forgot the story told by his mathematics teacher about a bricklayer who lifted with great effort heavy blocks of stone to the roof of a house. The work so performed remained stored up until the day the block was loosened and dropped on the head of an unsuspecting passerby. That simple story implanted vividly the full validity of the law of conservation of energy into Planck's mind.

Max Planck devoted himself to science to gain an insight into the impressions we receive from the world about us. Such a quest he regarded "as the most sublime scientific pursuit in life," and his introduction of the quantum of action fulfilled that aspiration. *(Photo purchased in Munich by the author)*

¶ After graduating from the Gymnasium, Planck attended the University in Munich for three years, and followed that with a year in Berlin. He studied experimental physics and mathematics because theoretical physics was not recognized as a discipline at that time. At Berlin his horizons were widened by the guidance of Herman Ludwig Ferdinand von Helmholtz (1821 to 1894) and Gustav Kirchhoff. The former was both a physicist and a physiologist. His discovery of the ophthalmoscope in 1851 permitted detailed studies of the living human eye. He also studied color and the problem of color blindness. The work of Kirchhoff regarding studies of the emissivities of bodies was discussed above.

¶ One can imagine the influence these men must have had on Planck. However, Planck confessed:

> *The lectures of these men netted me no perceptible gain. It was obvious that Helmholtz never prepared his lectures properly. He spoke haltingly . . . moreover, he repeatedly made mistakes in his calculations at the blackboard, and we had the impression that the class bored him at least as much as it did us.*

Kirchhoff, on the other hand, delivered a carefully prepared lecture, according to Planck, but, "it would sound like a memorized text, dry and monotonous."

¶ Planck quenched his thirst for knowledge by assiduous reading, including the work of Helmholtz and Kirchhoff. From this reading, he became acquainted with the work of Rudolf Julius Emmanuel Clausius (1822 to 1888), who had deduced a proof of the second law of thermodynamics from the hypothesis that heat will not pass *spontaneously* from a colder to a hotter body. Planck recognized that this hypothesis needed to be clarified, since heat conduction cannot be completely reversed by any means.

¶ Planck reasoned that a reversible process differs from an irreversible process solely on the nature of the initial and terminal states of the process and not on the manner in which the process develops. Moreover, the terminal state is in many respects more important than the original state, as if Nature "preferred" it. Planck reasoned that this "preference" is measured by Clausius's entropy and that in every natural process the sum of the entropies of all the bodies involved in the process increases. This became the basis for his doctoral thesis at the University of Munich, which he completed in 1879.

¶ The effect of this thesis on the physicists of the day, according to Planck, was "nil." He wrote, "None of my professors at the University had any understanding of its contents . . . Helmholtz probably did not even read my paper at all. Kirchhoff expressly disapproved of its contents . . . I did not succeed in reaching Clausius." Nevertheless, this effort became the basis for dramatic changes in our concept of physics.

¶ In 1885, Planck was appointed as Extraordinary Professor of Theoretical Physics at Keil. In 1892, he became Professor of Theoretical Physics at Berlin, where he remained for the rest of his life. The measurements of blackbody radiation undertaken by Otto Lummer (1860 to 1925) and Ernst Pringsheim (1859 to 1917) at the German Physico-Technical Institute were brought to Planck's attention, and he became interested in providing a theoretical foundation to describe their results. He began with the assumption that a cavity is filled with simple linear oscillators, each having a characteristic period. By simply substituting the energy of the oscillator for the energy of the radiation, Planck replaced a complicated structure having many degrees of freedom with a simple system having but one degree of freedom. However, the approach was unsuccessful.

¶ Planck approached the problem a second time, this time from the opposite side. He was now on "home territory," namely on the side of thermodynamics. Instead of correlating the temperature of the oscillator with the energy, Planck correlated the entropy. This approach had the distinct advantage of being noncompetitive with the work of others. Planck could now work "with absolute thoroughness, without fear of interference or competition."

¶ The results Planck reached were submitted to the Berlin Physical Society at its meeting on October 19, 1900. The paper was received well, for it corroborated the experimental work that others had done. However, Planck considered it as "lucky intuition" and devoted himself to the task of investing it with true physical meaning. He studied the interrelation of entropy and probability and found that by postulating the entropy to be proportional to the logarithm of the probability, he could properly interpret his radiation law. This was reported to the Physical Society in Berlin on December 14, 1900. This report introduced the concept that the ultimate energy of an oscillator is quantized, and the date is generally recognized as the birth date of the quantum theory.

¶ Although the significance of the quantum of action (as Planck called it) for the interrelation between entropy and probability was conclusively established, its role in physical processes was still an open question. Planck tried to incorporate it into the framework of classical physics, but he was unsuccessful. Planck did realize that the elementary quantum of action would play a significant role in physics, and he did recognize the need to introduce totally new methods of analysis and reasoning when dealing with atomic concepts. However, he could not formalize this.

¶ Albert Einstein (1879 to 1955) published three papers in the *Annalen der Physik* in 1905 which extended Planck's reasoning. In one of these, Einstein showed that light consists of photons, or quanta of light. One of the other papers presented Einstein's concept of relativity. Planck recognized that the Theory of Relativity confers an absolute meaning on the velocity of light, which has but a relative significance in classical theory. He reckoned that a parallel existed between the relationship of light velocity to relativity and the relationship of the quantum of action to blackbody emission.

¶ Planck was awarded the 1918 Nobel prize for physics. He endured many bereavements during his life. His eldest son, Karl, died at Verdun in 1916 during World War I. His second son, Erwin, was executed by the Nazis in January 1945 for his part in the plot on Hitler's life. Planck's house was completely lost in a fire during World War II. This loss included his library, which had been accumulated during his lifetime. He was also buried in the rubble for several hours after an air raid during that war. Nevertheless, he remained intellectually active and creative to the great age of eighty-nine, dying on October 3, 1947.

¶ Optics—and physics—were revolutionized by the efforts of this quiet man.   ◎

W. W. Coblentz, raised under primitive conditions on a farm in Ohio, rose to become a pioneer in infrared spectroscopy. *(Photo courtesy of National Bureau of Standards)*

The Blooming of American Science

William Weber Coblentz

THE PIONEERING SPIRIT with which America was founded endowed the country with a hardy people. Much of the success of the industry established here can be attributed to that spirit. So, too, has been the establishment of scientific stature in the United States. By the same token, the fact that the country was undeveloped slowed progress initially. Early contributions to science from America were sporadic. Notable exceptions to this include Joseph Henry (1797 to 1878) who gained fame for his experiments on electromagnetic induction, Josiah Willard Gibbs (1839 to 1903) who was responsible for much of the basic work in modern chemical thermodynamics, and Jean Louis Rodolphe Agassiz (1807 to 1873) who was a leading geologist, paleontologist, and zoologist. In fact, until a few generations ago, many American scientists went abroad for their advanced education. This began to change about the beginning of the twentieth century, and the change was brought about by true pioneers. The example of William Weber Coblentz (1873 to 1962) dramatizes this.

¶ Born in a log cabin on the family farm near North Lima, Ohio, on November 20, 1873, Coblentz began life without the "necessities" of today. He recalled that the farm "was not provided with a 'specialist's house' . . . where one could read the Montgomery Ward Catalogue." The farm dwellings consisted of a small log house which was occupied by Coblentz's father and family; and a large two-story brick house which had been built by William's grandfather in 1859. In addition, there was a carpentry shop, a barn, a cider press, a large smokehouse, and some miscellaneous buildings (many in need of repair).

¶ Like many of us, Coblentz had difficulty in tracing his ancestory back beyond his grandparents. He assumed that they had probably come from Coblentz on the Rhine and had migrated to Frederick County, Maryland, sometime prior to the American Revolution. His grandfather moved to Ohio in 1804 as a boy of twelve, and at the age of sixty-seven constructed the brick house William remembered.

¶ William's father, David, lived to an age of fifty-one, but his mother survived for only three years after William was born. His father remarried two years later, and William records his "highest regard for my second Mother."

¶ The Coblentzes moved frequently, David accepting employment as a share cropper. Coblentz described the district schools as lacking organization, "Thinking that they were really saving money on their taxes, the farmers hired the cheapest teacher." As a consequence, Coblentz felt that at the age of seventeen he had hardly received a grammar school preparation, "at least as regards book knowledge." But he was a perceptive lad and he gained considerable knowledge from observing the operations of the farm and the course of nature that existed in the open.

¶ Late in October 1891, Coblentz broke his home ties to attend the nearby Poland

Seminary, where he had decided to prepare for a career in electrical engineering. Considering leaving home, Coblentz reminisced, "The pathetic part was the struggle of my Father to raise $3.00 per week to pay for my board and place to stay in Poland." Much of the money had been raised by taking produce to Youngstown, "through wet, cold, and sleet" and selling it for very little profit. Upon arriving in Poland, Coblentz had "$4.38 in money . . . a light brown suit . . . a long-tailed coat . . . (and) two white shirts with bosoms starched stiff as a board."

¶ At the Seminary, Coblentz's curriculum consisted of courses in algebra, arithmetic, grammar, Latin, etc. Coblentz noted, "Just why I should be taking such courses when I was to become that nebulous something called an 'electrical engineer' was a staggering question to my Father, who was struggling to raise $3.00 per week. I frankly admitted that I did not understand why; but I explained to him that it was the recommended procedure, and the only course that I could pursue; also that after this preparation I could earn money by teaching."

¶ By the end of December, Coblentz was offered a job with the family physician, Dr. Seth H. Truesdale, "to take care of the horses, etc." This would pay for his boarding in the winter and supply him with $10.00 per month in the summer. Coblentz accepted this, retaining his job as janitor at the Seminary. For Dr. Truesdale, Coblentz had to care for four horses, milk two cows ("one of which would kick at the most unexpected moments"), feed the pigs and chickens, and cultivate the garden. He did this and arranged to work for others, for which he received $1.50 per day. By the end of the summer he had deposited $30.00 in the Poland bank.

¶ In the meager library of the Poland Seminary, Coblentz found an old copy of Avery's *Physics*, which had been discarded. He began to read that avidly as time permitted. However, life in the Truesdale home was not supplied with abundant leisure.

¶ One of the jobs he undertook for others consisted of whitewashing the cellar of the home of Mr. Isaac P. Sexton. This job affected the future course of Coblentz's life. Mrs. Sexton took an interest in Coblentz and brought him to work there with greater time for study. Mrs. Sexton, incidentally, was the daughter of the founder of the Poland Seminary. She inherited the inspirations of her father and emulated him by helping others, Coblentz being the most significant example.

¶ Coblentz now began to be molded into the world of civility. For the first time, he was given a warm place to bathe and warm water for that purpose. Inquiries were made regarding the care of his teeth. The crudeness of his speech was corrected. He was given clean towels and provided with suitable nightclothes. He became tractable in his new surroundings. It was soon impressed upon him that his foremost consideration was his education.

¶ The Seminary had once been equipped with physics apparatus. Coblentz found this equipment and cleaned and repaired what he could, using Avery's *Physics* as a guide. One of the instruments was a friction electric machine. Coblentz read that it was necessary to "ground" the instrument. He did this by lugging a coal bucket full of wet soil up two flights of stairs. He later realized that this was not the intent of grounding, "But then, in those days boys were not so smart; especially farm lads with less than seven months freedom from cleaning cow-stables and breeding pigs."

¶ Upon encouragement from Mrs. Sexton, Coblentz now set his goal on an electrical engineering education at Cornell. However, this required another year of high school preparation. He did this at Rayen High School in Youngstown during the academic year 1894/1895. He still lived in Poland, driving back and forth over the twelve and a half mile trip in a two-wheeled cart or on a bicycle over gravel roads.

¶ Graduating from high school in 1895, at the age of twenty-two, Coblentz anticipated entering Cornell the following year. However, financial reverses made this impossible. To this time he had not considered Case School of Applied Science, which had been established fifteen years earlier. Now it became appealing, particularly since the tuition was lower and there would be a savings in railroad fare. Moreover, passing the state scholastic examination would deduct $75.00 from his tuition.

¶ Having prepared himself for Cornell, he was ahead of his class at Case. Accordingly, he had free time on his hands which he used to take a math course at the adjoining Adelbert College. Thus, by the end of the first semester Coblentz had practically no classwork. He returned to Poland where he spent the winter working out the problems in a book on differential calculus.

¶ In the spring, financial matters looked brighter and Coblentz enrolled at Cornell. But a few weeks later he learned that the financial picture had been reversed, so he was forced to return to Case. Soon thereafter he changed his major to physics as that would permit him to enroll in a course in astronomy. In June 1900 he was graduated "B.S. majoring in physics."

¶ In September 1900, Coblentz realized his ambition to study at Cornell. Although the entrance requirements at Case were less stringent than at Cornell, it is obvious that Coblentz did not receive a second-class education. In June 1901, he was awarded an M.S. degree from Cornell, and two years later he received his Ph.D.

¶ At Cornell, he began researches into the characteristics of the spectra of various substances. This research was of great importance, previous spectra being somewhat crude, and very few extending to 14 μm. Much of the apparatus Coblentz used had to be built. Certainly, his farm experience repairing machinery stood him in good stead when he constructed his scientific apparatus.

¶ Quantitative measurements of infrared spectra of pure molecular compounds appeared to be a fruitful area for further investigation, and Coblentz was made a Research Associate of the Carnegie Institute at a salary of $1000 a year. Coblentz remained in Ithaca for two more years systematically mapping the infrared spectra of thousands of molecular substances.

¶ In the spring of 1905, Coblentz joined the recently formed National Bureau of Standards as a laboratory assistant at a salary of $900 a year. For the next forty years, Coblentz's contributions at the Bureau became practically legendary. He continued his work on the infrared spectroscopy of various molecules. He became the first to accurately determine the spectral radiation from a blackbody, thus verifying Planck's law experimentally. He was one of the first to measure the heat radiated from stars, planets, and nebulae. The thermopiles he constructed were sought after because of their exceptional characteristics.

¶ Coblentz was honest and straightforward—the epitome of a true pioneer. In later years he realized that it is sometimes necessary to be politic, but he always felt, "the most successful way is to speak out frankly and hit hard." Speaking of hitting hard, when he was a sophomore, a senior who was following Coblentz up a flight of stairs pulled one of Coblentz's feet while the other was lifted. Coblentz fell on the stone steps with both knees. He was hurt, but he jumped up and struck the senior with a crushing blow on the side of the face. The prankster learned a hard lesson, for he suffered a lacerated jaw.

¶ During his first year at Cornell, Mrs. Sexton paid for part of Coblentz's costs. During his last two years, he obtained a scholarship that paid $300 a year, and he managed on that. When the Carnegie money was realized, Coblentz continued to live in the same style, and repaid what he owed with the balance. But his appreciation to Mrs. Sexton went far beyond merely repaying her financially. In 1904, Mrs. Sexton fell on the ice and received double fractures in the wrists. These were not set properly, leaving her almost helpless. She also suffered with arthritis and spinal pain. In 1905, Coblentz brought her to Washington and cared for her until she died in 1923.

¶ Coblentz had a great interest in the education of young men of potential. He endowed a scholarship fund at Case, doubtless recalling his own difficulties and desiring to ease the financial burdens of others.

¶ A curious aspect of Coblentz's character was his interest in spiritualism. It began with childhood experiences which could not be explained by the logic of science. In adult life he frequently went to seances, often with scientific equipment to help him either to understand the phenomena or to debunk the medium. I once asked one of Coblentz's colleagues if his interest in spiritualism affected his science. I was told emphatically that it did not. Coblentz would watch a galvanometer deflection and his experience would indicate the quality of the measurement. There was no spirit voice telling Coblentz whether or not his measurements were valid!

¶ Coblentz's health was poor. He was troubled with tuberculosis most of his life, possibly as a result of sleeping in a drafty environment as a child. He had digestive disorders which forced him to go on a strict diet. It is a measure of his character that he contributed so much in spite of his handicaps.

¶ Coblentz's scientific contributions are of significance to botany, physiology, and psychology as well as to physics and optical engineering. The thoroughness and accuracy of his work established great confidence in him among his peers.

¶ It has been said that the adoption of radiometric standards from the extreme ultraviolet to the far infrared is the result of his efforts.

¶ William Weber Coblentz was a true pioneer.

Wartime Incentive

Robert Joseph Cashman

LTHOUGH ONE CANNOT ADVOCATE WARS to further the advance of science, there can be gains realized during the pressures of hostilities that are beneficial to mankind. The development of infrared sensitive detectors during World War II is a case in point. Research on photosensitive cells which was conducted by the adversaries enabled scientists in the succeeding peace to undertake investigations that have advanced astronomy, geology, medicine, agriculture, meteorology, and other scientific disciplines.

¶ The photoconductive effect was first reported by Willoughby Smith in 1873. He had been experimenting with selenium as an insulator for submarine cables and noted that crystalline selenium offers considerably less electrical resistance when exposed to light than when it is kept in the dark. Since the response of selenium extends slightly into the infrared, a few scientists reported using this material for investigations requiring an infrared sensitivity as early as 1904. However, such cells were unreliable and measurements using these made little impact.

¶ By the mid-1930s reliable photocells having photoemissive surfaces were being manufactured. Accordingly, interest in photovoltaic and photoconductive cells was directed primarily to investigations of material behavior. As World War II approached, the desire to "see" in the dark and to communicate unobtrusively goaded the military services (in Germany, the United States, and England) to support efforts to develop improved photoconductors with a sensitivity in the infrared.

¶ The first important photoconductive detector with a significant infrared response was lead sulfide. Following the successes achieved with this material, investigations of other lead salts began. This activity has continued because of the usefulness of infrared investigations in many fields of endeavor.

¶ The events in Europe prior to 1940 revealed clearly to Harvard President James Bryant Conant, M.I.T. Vice President Vannevar Bush, and others, that American science must be organized to meet the impending threat, and such a recommendation was made to President Franklin Roosevelt. As a result, the National Defense Research Committee (NRDC) was formed in June 1940. A year later, the Office of Scientific Research and Development (OSRD) was established with NRDC as one of its units. The impact of this organization has had a profound effect on the course of history. It has been asserted that although the hardware developed (atom bombs, radar, sonar, antibiotics, etc.) changed history, the greatest invention was political: a new way of using brainpower.

¶ Late in 1942, an Optics Division of NRDC was formed to carry out a broad program which included studies of camouflage principles, aerial mapping, aircraft detection by infrared radiation, communication using infrared signals, and the development of the sniperscope. The need for improved infrared detectors was obvious, and detector research was undertaken from a variety of directions.

¶ Wartime research in the United States on photoconductors proceeded initially with investigations of silicon and thallous sulfide cells. The latter were found to be superior, but did not represent a useful detector for military applications. During the first World War, T. W. Case had experimented with this type of detector. He had successfully detected messages sent over a distance of eighteen miles, but his detectors were unreliable. In particular, they suffered fatigue when exposed to short wavelength radiation. Neither was it possible to manufacture cells with reproducible results. With the war's end in 1918, interest in further development ceased.

¶ On December 1, 1941, Northwestern University Physics Professor R. J. Cashman was asked by NRDC to investigate infrared sensitive cells. Cashman's investigations brought a significant change to the infrared community. Robert Joseph Cashman was born on September 27, 1906, at Wilmington, Ohio. As a youngster, Cashman made selenium cells which he used to perform simple experiments. Partially for this reason he studied physics in college, receiving his bachelor's degree from Bethany College in 1929. He then transferred to Northwestern University, where he obtained an A.M. degree in 1930 and a Ph.D. in 1935. He then joined the physics faculty there.

¶ His early interest in devices which convert photon energy to electrical energy was enhanced by his formal training, and his subsequent research was in this direction. By 1939, he began investigations of the photoemissive characteristics of semiconductors in hopes of enhancing an understanding of solid state mechanisms. During these investigations, Cashman noted the photoconductive effects, but these were subsidiary to his main interests and he did not pursue this at that time.

¶ When Cashman was asked by NRDC to extend his investigations to infrared sensitive devices, he began with Tl_2S, evaporating the material over a grid on the inner surface of an evacuated tube. By diligent experimentation, he found that the necessary ingredient to sensitize the cells is oxygen. By late 1943, Cashman reached the point where Tl_2S cells of reliable characteristics could repetitively be made. This permitted him to turn the process over to mass manufacturing techniques and he was freed to investigate the properties of other materials.

¶ Cashman began with investigations of MoS_2 and Ag_2S; this was expanded to include PbS in February 1944. The potential of MoS_2 appeared to be the most promising as a result of using natural crystals. However, when investigations of synthetic crystals were begun, it was clear that PbS would make the best infrared detector.

¶ Most of the sensitization experiments Cashman performed were purely empirical as there was but meager theoretical understanding to guide his investigations. He subjected chemically prepared layers to heat treatments in air and in oxygen at a variety of pressures. He also investigated treatment in a vacuum, which surprisingly (since he had discovered that oxygen is necessary to energize the material) resulted in the best signals. Cashman measured time constants, resistances, spectral dependence, noise, applied voltage effects, and other parameters as well as the sensitivity for a wide variety of conditions which he used in preparing cells. This was a big stride toward elucidating the mechanisms of photoconductivity, albeit a time-consuming effort.

¶ In December 1944, Cashman was sent some captured German photoconductive cells. His studies showed that they were PbS. Historically, the German investigations

A MoS$_2$ cell built by W. W. Coblentz before 1918, and supplied to Robert Cashman in his World War II investigations. *(Photographed by the author at Cashman's laboratory in Evanston, Illinois)*

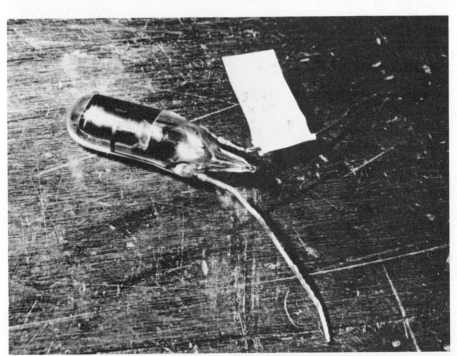

An early Cashman PbS cell, c. 1945. *(Photographed by the author at Cashman's Northwestern University Laboratory in Evanston, Illinois)*

of lead salt detectors significantly antedate those undertaken in the United States. However, their impact was felt only after the independent discoveries of Cashman. Naturally, the German military restrictions had prevented information of their activity from being disseminated in Allied countries.

¶ One of the German pioneers in infrared detector development was Edgar W. Kutzscher, who in 1930 began research at the Department of Physics at the University of Berlin. He was then interested in the electrical and physical characteristics of crystal rectifiers, and found that their resistance changed when exposed to visible or infrared radiation. In the course of subsequent investigations, Kutzscher and a collaborator discovered that PbS had a superior sensitivity to infrared radiation than any other radiation detector then known.

¶ Beginning in about 1933, Kutzscher's research was supported by the German Army. The original hope for the utilization of infrared sensors was to assist radar since these devices had too broad a field to precisely locate an object angularly. When radar was improved to attain angular resolution, interest in infrared faded; the realization that radar can be jammed renewed the interest in infrared.

¶ Considerable research and manufacturing of infrared detectors was done at the Electroacustic Company in Kiel. However, it was realized during the Allied bombing of Germany that this site, located close to the North Sea, was vulnerable. Hence, it was moved to Namslau in eastern Germany late in the war. In January 1945, fears that the Russians would capture this area prompted the Germans to consider moving once again. However, the Russian advance was too fast to permit the plan to be executed. On January 25, 1945, the Russians occupied the factory and obtained possession of most of the records and equipment. It was also at about this time that Cashman received the captured PbS cells.

¶ With the declaration of peace, there was a dramatic decrease internationally in the interest in detector development. However, within a few years, this interest was revived. Infrared radiation is now an important element in investigations of phenomena pertinent to a number of disciplines. The peaceful applications of infrared technology can be traced, in large part, to the investigations made during a period of intense hostility. One should also note that most of the developments of infrared detector technology can be traced to the pioneering efforts of one man: Robert J. Cashman. ⌀

Coating Telescope Mirrors

John Donovan Strong

 CIENCE NEEDS IMPROVED TECHNOLOGY to provide instrumentation to gather data from which insights can be made to advance science. Technology, by the same token, requires advances in science before technological improvements can be scored. If this sounds like one is trying to lift oneself with one's bootstraps, consider the events leading to the coating of the Palomar telescope mirror. To do this, we must briefly review the career of John Donovan Strong (b. January 15, 1905).

¶ Strong, a native of Kansas, obtained his A. B. degree at Kansas in 1926 and then entered the University of Michigan where he studied physics. Strong's interest was in infrared spectroscopy, and he was eager to explore spectral regions theretofore unknown. By growing large alkali crystals, he was able to form prisms that he used to extend the available spectral range from about 15 μm to beyond 30 μm. The arrangement he preferred for his spectrometer was the Littrow type, which requires that radiation be reflected through the prism a second time. However, it was impossible to chemically deposit silver on a KBr prism since it is hygroscopic and the chemical process is wet. Moreover, a silvered flat used in conjunction with the prism leads to faint double images and interference bands. Imagine, therefore, Strong's reaction to a colloquium presented in 1929 by Professor L. S. Ornstein, from Holland. He described a technique developed in Europe in which a silver wire placed in a tungsten helix was heated in a high vacuum where it evaporated and condensed as a mirror on adjacent surfaces.

¶ Realizing that this technique was the answer to achieve the silvering of his prism face, Strong petitioned his aegis professor, Harrison McAllister Randall (1870 to 1969), to repeat Ornstein's procedure. However, Randall could not permit this as another Michigan professor had indicated a desire to carry out this work. But, after three months had elapsed (which seems an eternity to a graduate student) with no activity by that professor, Strong was given the go-ahead. His KBr prism was soon properly silvered, and shortly thereafter Strong completed his research, obtaining his Ph.D. in 1930.

¶ Following graduation, Strong was appointed as a National Research Fellow at the California Institute of Technology. There, he teamed with C. Hawley Cartwright to extend his vacuum deposition efforts by investigating the evaporation of various materials in high vacua. The metallic elements they investigated included Al, Sb, Be, Bi, Ca, Cr, Co, Cu, Au, Fe, Pb, Mg, Mn, Ni, Se, Ag, Te, Sn, and Zn. They also investigated the possibility of evaporating alloys, finding that speculum metal behaves rather well, probably because tin and copper have roughly the same vapor pressure at the evaporating temperature. On the other hand, brass was found to distill fractionally, the zinc coming down first. Jeweler's silver was also found to distill fractionally, the silver being evaporated before the copper. Non-metals with which they experimented included quartz, fluorite, the alkali halides, and silver chloride. Glass was found to be fractionally distilled, leaving a residue resembling opal.

¶ The technology of vacuum deposition was, of course, in its infancy and required research to identify the chemical reactions taking place. In addition to investigating the evaporation of various materials, consideration was given to the heating processes. Experiments with tungsten helix and tungsten coils were performed. Crucibles of graphite, pure fused magnesia, and alumina (sapphire) were also used as was thorium oxide. Experiments were performed to determine the effects of substrate temperature on the tenacity of the deposition. Thus, we see the need of scientific inquiry in developing a technology.

¶ But perhaps the impact this technology made on science is more significant than the impact science made on the technique of vacuum depositions. By January 1931, Strong was able to deposit silver on an astronomical telescope mirror. He did this for Professor Philip S. Fogg, the Registrar at CalTech, who had ground and figured a six-inch mirror. Strong protected the silver with an overcoat of quartz. Although a mirror surface prepared by vacuum deposition is superior to one prepared by chemical deposition, aluminum offers even greater advantages over silver. These include a superior reflectance at short wavelengths, permanance, and an ease in cleaning. In 1932, Strong mastered a technique to evaporate aluminum. It is interesting to note that Strong re-coated Fogg's mirror and that this mirror (which is the first astronomical mirror to have been coated by evaporated aluminum) is now in Strong's laboratory in good condition after nearly a half century.

¶ The astronomical community responded quickly and enthusiastically to the success of Strong's evaporated aluminum coatings and he was asked to coat many mirrors. These included the 15-inch mirror at the Lowell Observatory, the 24-inch reflector at

Standing alongside the primary mirror of the 200-inch Mt. Palomar telescope in 1947 are John Strong, who was responsible for coating the mirror, & Don Hendrix, who figured the mirror. *(Photo courtesy California Institute of Technology)*

the Yerkes Observatory, and many more. The superiority of evaporated aluminum had a profound influence on astronomy. For instance, Strong found that at the Cassegrainian focus of the 60-inch Mt. Wilson telescope, the limiting magnitude obtained with evaporated aluminum was equivalent to that achieved with the 100-inch instrument if coated with silver.

¶ In the early 1930s, the concept of the 200-inch telescope was gestating. Consideration was being given to silver, chromium, and aluminum to coat the objective, plus the process of deposition. Strong was involved in these deliberations, but his fellowship was due to expire in September 1932. As the date approached, he was filled with trepidation, for these were years of the depression. Fortunately, the need to perform a variety of tests in anticipation of the 200-inch telescope prompted Robert A. Millikan (1868 to 1953), President of CalTech, to appoint Strong as a Research Fellow in Astrophysics to work on the Palomar project at a salary of $2000 per year—a salary Strong considered a bonanza!

¶ Strong made reflectivity measurements of various metallic depositions. He studied the problems associated with recoating. He considered problems associated with scattering. All this was done during the period he was aluminizing the mirrors previously mentioned. It soon became clear that evaporated aluminum was the route to go with the Palomar mirror. But placing a 200-inch mirror inside a bell jar is no mean problem. A chamber to house this mirror during deposition had to be constructed.

¶ The vacuum chamber Strong built had a 19-foot inside diameter and a 7-foot inside height. It weighed thirty tons. The aluminum was evaporated from 175 tungsten coils located in a plane twenty inches from the face of the mirror. By 1941, the vacuum chamber was nearly completed, but wartime activity interrupted further progress. Strong went to Harvard where he performed important research regarding the detection of military targets with infrared technology. After the war, Strong joined the faculty of the Johns Hopkins University. However, he was loaned to the Palomar Observatory in the summer of 1947 to complete the coating of the 200-inch mirror, which he finished by Thanksgiving of that year.

¶ The impact on astronomy by the technology which Strong created is obvious. Astronomical research requires the collection of a copious supply of photons. This demands both large mirrors and a highly reflecting surface. But the contributions of vacuum depositions extend far beyond providing high reflectances. Interference filters and non-reflecting surfaces are produced with vacuum depositions. Microcircuitry depends upon this technology. Vacuum deposited films are used as efficient polarizers with large apertures. Figuring of aspheric mirrors can be accomplished with the aid of evaporation techniques. The impact of vacuum deposition technology is felt in many scientific disciplines. And that technology was accomplished through rigorous scientific study.

¶ John Strong is to be acclaimed not only for his contributions to technology. His researches in the fields of astrophysics, meteorology, infrared spectroscopy, and the optical properties of materials are legendary. His former students form a cadre of exceptional contributors to optics. In many respects he exemplifies the interrelationship of science with technology.

Holography—The New Photography

Joseph Niepce, Louis Daguerre, William Fox Talbot, George Eastman, Dennis Gabor, Emmett Norman Leith, Juris Upatnieks

ADVANCES IN OPTICS, as in most branches of science, result from achievements in technology which make possible more refined observations. Although the converse is a bit rarer, advances in optical technology are also achieved by an improved understanding of the nature of light. An outstanding example of this is the invention of holography. Although it represents a form of photography, holography constitutes a distinct departure from the traditional practice of pictorial recording. It achieves this departure by being based upon a knowledge of the constitution of light. It is of interest to review the differences.

¶ Image forming by the camera obscura was known in ancient times. The blackening effect of light upon silver salts was known to the alchemists. A German chemist, Johann Heinrich Schulze (1687 to 1744), used this property to imprint shadow pictures on a salt precipitate in a bottle. The blackening effect of the blue light in prismatically dispersed sunlight was found to be more effective than the red in this regard by Carl Wilhelm Scheele (1742 to 1786). Johann Wilhelm Ritter (1776 to 1810) extended this discovery to ultraviolet rays. However, at the start of the nineteenth century, photochemistry had yet to advance sufficiently to permit the discovery of the photographic process.

¶ About 1814, Joseph Nicephore Niepce (1765 to 1833) undertook to automatically transfer pictures to a lithographic block. By 1827, he succeeded in making permanent records. These depended upon the action of light to reduce the oil solubility of a preparation of asphalt and lavender oil, spread upon a plate of silver or glass instead of upon the stone then customarily used in lithography. Exposures of nearly a day would be required of this process if a lens was used. Niepce became associated with Louis Jacques Daguerre (1787 to 1851), a painter who specialized in lighting effects to attain a realistic three-dimensional effect in his work. They strove without apparent success to perfect Niepce's discovery. Some time after Niepce's death, however, Daguerre did perfect a technique to improve the process by exposing the plates to mercury vapor. The Daguerreotype process became an immediate success.

¶ During the period that Niepce and Daguerre were experimenting with their process, experiments on photochemistry were progressing in England. John Frederick William Herschel (1792 to 1871) discovered the fixing properties of sodium thiosulfate in 1819. Twenty years later, he exhibited pictures made with this process and coined the word "photograph." In this period, William Henry Fox Talbot (1800 to 1877) had been using a camera obscura to make sketches, which led him to investigate means of fixing images by chemical means. By 1835 he had been able to record images on paper coated with silver chloride and fixed with sodium chloride. Thus began the negative/positive process of photography with which we are familiar today.

¶ During the next half century, photography was practiced by professionals and adept amateurs. The first model of a cheap box camera for the mass market was produced by George Eastman (1854 to 1932) in 1888. Today, we are familiar with numerous cameras and the broad market that they reach.

¶ The photographic process is essentially one of recording the two-dimensional distribution of light energy in the image of a given scene. It occurred to Dennis Gabor (1900 to 1979) in 1948 that the wavefront emanating from each point of a scene could be recorded by causing it to interfere with a background wave, thus converting phase difference into an intensity difference. Moreover, the wavefront could be reconstructed by illuminating the recorded information with coherent light. Gabor's objective in this effort was to record the wavefront generated in an electron microscope, and to form the reconstruction with visible radiation. In this research, Gabor used a mercury lamp to reconstruct the image, hoping to gain additional magnification by scaling from electron wavelengths to that of the mercury lamp. Gabor termed this process *holography*, meaning whole record.

¶ Although others soon sought to perfect Gabor's invention, the numerous practical difficulties encountered prevented attaining the desired imagery. The result was a virtual impasse in further development. Fortunately, wavefront reconstruction was independently discovered by Emmett N. Leith while investigating optical processing techniques to record and display radar signals. This soon led to a practical and useful holographic process.

¶ Emmett Norman Leith was born on March 12, 1927, at Detroit, Michigan. He obtained B.S. and M.S. degrees in physics and a Ph.D. degree in electrical engineering at Wayne State University. Upon joining the staff of the University of Michigan, he became associated with the group interested in improving the recording and display of radar signals using optical processing. It soon occurred to Leith that if the radar signal received along a real (or synthetic) aperture is stored on a transparency and then illuminated with coherent light, the emerging wavefront will duplicate the radar wavefront. Leith realized, of course, that the reconstructed field would be a miniaturization of the original field, due to scaling laws associated with transforming the microowave frequencies to optical frequencies. But, Leith reasoned, one could now "see" the radar image.

¶ Leith submitted a report to his superiors at Michigan on May 22, 1956, in which he proposed a data processing system which could be considered an optical analog of a radar system. In that report, Leith described how to avoid the twin images associated with the diffraction into positive and negative orders. He further described in this report a means to reconstruct the wavefront with white light. These were significant descriptions because they permitted holography to "work."

¶ Although Gabor's work had been published prior to the time of Leith's report, the latter was unaware of that effort. It could hardly be expected that engineers working with radar systems would be conversant with articles in *Nature*, where Gabor's results were published. Moreover, Gabor's effort was not widely known, as the procedure was masked with experimental problems that prevented it from becoming an acceptable tool. Gabor had hoped to develop a process whereby aberrations associated with elec-

Emmett Leith and Juris Upatnieks preparing a hologram in the laboratory. *(Photo courtesy of the University of Michigan)*

tron microscope imagery could be corrected. This had not proved to be feasible with Gabor's approach, principally because of the overlapping images formed in each order.

¶ The key to Leith's success lay in the fact that the amplitude and phase of microwave frequency signals had been studied thoroughly and were routinely recorded. Gabor, of course, had to struggle with the much higher optical frequencies. At mircowave frequencies, information about the wavefront can be ascertained instantaneously, whereas at optical frequencies, one could then only obtain time averages.

¶ When Gabor's work was brought to Leith's attention, it was received with mixed emotions. On one hand, Leith was disappointed not to have been "first." On the other hand, it gave him greater confidence in the merit of his approach. Radar engineers, being unaccustomed to the nomenclature of optics, had not responded with enthusiasm to Leith's approach. However, with the renewed confidence gained by the knowledge of Gabor's work, Leith managed to change the engineers' attitudes.

¶ Once Leith's process had been demonstrated as practical for radar signal processing, interest was revived in all-optical wavefront reconstruction. Together with his colleague, Juris Upatnieks, Leith began experiments of recording optical wavefronts. One of the major advances soon made was the adoption of a laser as the source of illumination. For radar signal processing, the coherence length of the microwave radiation is sufficient to form the hologram. This can be reconstructed with a mercury lamp since the hologram has no depth. However, for a three-dimensional object, it is necessary to use a source with a long coherence length to record the hologram; a laser provides this.

¶ With Leith's successes, the field of holography blossomed. The efforts of Leith and Upatnieks stirred the imaginations of physicists and engineers. Following the announcements of the achievements of Leith and Upatnieks in 1961 and 1962, legions of scientists began investigations of holography. It soon became applied to many scientific and technical problems. Today, holography is practically a subfield of physics in its own right.

¶ Holography, in essence, is a recording and reconstruction of a wavefront. It thus differs from photography, and the purist may object to relating the two. However, they are both concerned with recording an image, so I feel justified in discussing them together. The reconstructed hologram wavefront is identical to that which issued from the object; hence, a complete three-dimensional image is restored. Naturally, this could not have been done in ignorance of the wave character of light. Holography may thus be considered as an advance from conventional photography in image recording which has resulted from an improved understanding of the nature of light.

Retrospection

N ADDITION TO EXAMINING THE OBSTACLES man faced in his search to determine the nature of light, one naturally inquires if some pattern exists whereby we can identify the men who made significant contributions to optics. Did they do their work in their younger years? Did they generally have extensive educations? Were their lives free (or full) of outside pressures such as political entanglements, religious bigotry, or pestilence? Were they liberal or conservative in their outlook? Were they emotionally stable? Indeed, can one categorize these men who made substantial improvements in our knowledge of optics by any other attribute?

¶ We begin our inquiry by seeking the age at which the great names in optics made their contributions. Isaac Newton made his first (and possibly his greatest) investigations of light by the time he was twenty. However, Ibn al-Haitham accomplished his remarkable discoveries after he had reached the age of fifty. W. W. Coblentz began his contributions only when he had reached an age of nearly thirty, but he continued his work zealously beyond the age at which many of us retire. John Strong is another "great" who has continued to make substantial contributions throughout his life. I can only conclude that age has little to do with creativity, although one is often less encumbered by administrative and other interferences in younger years.

¶ We can quickly dispense with wealth. Certainly, it was an advantage to Huygens, Tycho Brahe, Galileo, Descartes, Planck, and others. However, Michelson, Coblentz, Fresnel, Fraunhofer, and others succeeded without this advantage. Wealth or class distinction does not appear to have had a discernible effect on those who contributed to optics.

¶ Perhaps a consideration of humility will provide us with an insight into the generalized character of the optics greats. William Herschel became embroiled in controversy, but it was not of his choosing and there is no evidence that his responses were contentious. Young was enmeshed in a similar dispute; apparently his response (although probably justifiable) did not defuse the situation. Newton, too, was involved in heated disputes which suggest that he was not even-tempered. Coblentz had a temper with a short fuse, but he is remembered as a gentleman. Tycho Brahe was an irascible tyrant. On the other hand, Emmett Leith is quiet, humble, and affable. There appears to be no connection between one's ability to contribute to optics and one's emotional reserve.

¶ Considerations of education, upbringing, and/or environment also are filled with contradictions, leaving us with little on which to categorize the men of optics. We could even consider their marriages. However, since some remained single, some married, and some were divorced, it is difficult to draw any conclusions in this regard.

¶ One should also examine the socio-political viewpoints of the greats in optics to see if a consensus exists. Many were liberal. Arago was responsible for freeing the French-held African slaves, for instance. However, others held a more conservative

Marcel Minnaert sketches some manifestations of light and color in the open air of Michigan. *(Photo by the author at the author's home [then] in Michigan.)*

outlook, so it would require a biased viewpoint to suggest that liberalism is associated with these men. Similarly, an attempt to relate a religious attitude with these men would be suspect.

¶ By and large, from this examination, we can conclude only that the men of optics represent a broad cross section of mankind. However, before we dismiss the subject with a terse statement to the effect that there is nothing to distinguish the optics greats from other mortals, let us consider their aesthetic attitude. Probably, this can be done only by conjecture. Certainly, no one wrote that he investigated optical phenomena simply to earn a living. On the other hand, a number of scientists have described their delight in examining optical phenomena. Melloni wrote about his experiences as a young man when he climbed a nearby hill to watch the sun rise and nature awake. This instilled in him a desire to probe more deeply into the nature of light. Planck described his motivation to study science and to use the knowledge gained to enable him to better understand the meaning of his sensory perceptions. It is difficult for me to conceive of an investigator of optical phenomena not being motivated by the emotional delights associated with these observations.

¶ One of the proponents of observing the beauty in nature was Marcel Minnaert, whose book *The Nature of Light and Colour in the Open Air*, has been an inspiration to many of us. In the preface of that book, Minnaert states simply that which I have been stumbling to enunciate: "A lover of Nature responds to her phenomena as naturally as he breathes and lives." Later, he writes, "Each moment he is struck by new and interesting occurrences. With buoyant step he wanders over the countryside, eyes and ears alert, sensitive to the subtle influences that surround him, inhaling deeply the scented air, aware of every change of temperature, here and there lightly touching a shrub to feel in closer contact with the things of the earth, a human being supremely conscious of the fullness of life."

¶ Minnaert suggested that it is wrong to think that viewing the "poetry of Nature's moods" in a scientific manner will lessen the splendor of the event. Rather, he states that "the habit of observation refines our sense of beauty." In my limited experience I have found this to be true. Looking for the "green flash" from the setting sun is a memorable experience. It was Minnaert who alerted me to it.

¶ Probably my bias prevents me from examining the lives of the great men of optics with complete objectivity. Nevertheless, I contend that one of the features that clearly distinguishes these men from others is the motivation to seek and understand the beauty associated with viewing nature.

¶ At least, I hope that it is.

Bibliography

Bernal, J. D. *Science in History*. C. A. Watts & Co. Ltd. , London, 1965.

Biographies of Distinguished Scientific Men. Books for Libraries Press, Freeport, New York. First published 1859; reprinted 1972.
Biographies of Arago, Herschel, Fourier, Malus, Fresnel, Young, and others.

Brewster, David. *Letters of Euler on Different Subjects in Natural Philosophy, Addressed to a German Princess, With Notes, and a Life of Euler*. Printed and published by J. & J. Harper, New York City, 1833.

Bugge, Thomas. *Science in France in the Revolutionary Era*. M. I. T. Press, Cambridge, 1969.

Coblentz, William Weber. *From the Life of a Researcher*. Philosophical Library, New York, 1951.

Crosland, Maurice. *The Society of Arcueil*. Harvard University Press, 1967.

Gamow, George. *Thirty Years that Shook Physics*. Doubleday & Co., Garden City, New York, 1966.

Gordon, M. M. *The Home Life of Sir David Brewster*. Edmonston and Douglas, Edinburgh, 1869.

Guillemin, Victor. *The Story of Quantum Mechanics*. Charles Scribner's Sons, New York, 1968.

Kuhn, Thomas S. *Black-body Theory and the Quantum Discontinuity 1894–1912*. Harvard University Press, 1967.

Lenard, Phillip. *Great Men of Science*. G. Bell and Sons, Ltd., London, 1958.

Lindberg, David C. *Theories of Vision from Al-Kindi to Kepler*. University of Chicago Press, 1976.

Livingston, Dorothy Michelson. *The Master of Light*. Charles Scribner's Sons, New York, 1973.

McGucken, William. *Nineteenth-Century Spectroscopy*. The Johns Hopkins University Press, Baltimore, 1969.

Bibliography

Minnaert, Marcel. *The Nature of Light and Colour in the Open Air*. Originally published by G. Bell & Sons, Ltd., London. Republished by Dover Publications, New York 1954.

More, Louis Trenchard. *Isaac Newton*. Charles Scribner's Sons, New York, 1934.

Ornstein, Martha. *The Role of Scientific Societies in the Seventeenth Century*. University of Chicago Press, 1928.

Planck, Max. *Scientific Autobiography*. Philosophical Library, New York, 1949.

Ronchi, Vasco. *The Nature of Light*. Harvard University Press, 1970.

Sabra, A. I. *Theories of Light from Descartes to Newton*. Oldbourne, London, 1967.

Sime, James. *William Herschel and His Work*. Charles Scribner's Sons, New York, 1900.

Singer, Charles, et al., Eds. *A History of Technology*. Five volumes. Oxford University Press, London, 1954.

Steffens, Henry John. *The Development of Newtonian Optics in England*. Science History Publications/USA, New York, 1977.

Taton, Rene, Ed. *History of Science*. Three volumes. Basic Books, Inc., New York, 1963.

Williams, Trevor I., Ed. *A Biographical Dictionary of Scientists*. Wiley-Interscience, New York, 1969.

Wood, Alexander. *Thomas Young Natural Philosopher*. Cambridge University Press, Cambridge, 1954.

Index

Page numbers in **bold type** refer to illustrations

OPTICAL ANECDOTES

This book was edited, designed, indexed, and typeset by Elaine C. Cherry.